Green Energy and Technology

Climate change, environmental impact and the limited natural resources urge scientific research and novel technical solutions. The monograph series Green Energy and Technology serves as a publishing platform for scientific and technological approaches to “green”—i.e. environmentally friendly and sustainable—technologies. While a focus lies on energy and power supply, it also covers "green" solutions in industrial engineering and engineering design. Green Energy and Technology addresses researchers, advanced students, technical consultants as well as decision makers in industries and politics. Hence, the level of presentation spans from instructional to highly technical.

Indexed in Scopus.

Indexed in Ei Compendex.

Zhijun Peng
Editor

Technological Advancements and Future Directions in Green Energy

Selected Papers from ICGET 2024

Springer

Editor
Zhijun Peng
University of Lincoln
Lincoln, Lincolnshire, UK

ISSN 1865-3529 ISSN 1865-3537 (electronic)
Green Energy and Technology
ISBN 978-3-031-83202-4 ISBN 978-3-031-83203-1 (eBook)
https://doi.org/10.1007/978-3-031-83203-1

This Springer imprint is published by the registered company Springer Nature Switzerland AG
The registered company address is: Gewerbestrasse 11, 6330 Cham, Switzerland

Preface

It is our great pleasure to welcome you to 2024 9th International Conference on Green Energy Technologies (ICGET 2024), which was held during July 21–23, 2024 in Berlin, Germany. ICGET 2024 is organized by the Frankfurt University of Applied Sciences.

The conference serves to foster communication among researchers and practitioners working in a wide variety of scientific areas with a common interest. The conference is also an excellent forum for scientists and engineers throughout the world to present and discuss the latest technological advancement as well as future directions and trends in the field. Previous ICGET conferences have been held in various cities including Frankfurt, Rome, Amsterdam, Kuala Lumpur, etc.

The program features three prestigious keynote talks delivered by excellent experts: Prof. Zhe Chen from Aalborg University, Denmark; Prof. Zhijun Peng from University of Lincoln, UK; and Prof. Dimitrios Karamanis from University of Patras, Greece. I would like to express my sincere gratitude for the insightful and inspiring keynote speeches. Their talks were not only engaging but also deeply enriching. ICGET 2024 has also attracted participants from all over the world, including Australia, Portugal, Sweden, Spain, South Korea, China, Ecuador, India, the Philippines, and Israel. We had the privilege of listening to insightful and engaging presentations, and everyone enjoyed sharing and discussing their perspectives during the conference. I believe this experience will prove to be highly rewarding for all attendees.

We thank the authors of all submitted papers and invited speakers for providing the content of the program. We would like to express our gratitude to the program committee, who contributed their expertise, provided valuable feedback to the authors, and helped improve the quality of the accepted papers.

Many members of the organizing team worked very hard to turn our initial visions for this conference into reality, we would like to warmly thank all organizing committee members for their dedication before and after this unique event. Your expertise, enthusiasm, and time commitment enabled us to prepare the final program.

We hope that all participants and other interested readers benefit from and enjoy the presentations and proceedings and also find it stimulating in this process.

Lincoln, Lincolnshire, UK Zhijun Peng

Committees

Conference General Chair

Hartmut Hinz, University of Applied Sciences, Frankfurt, Germany

Conference Chair

Zhijun Peng, University of Lincoln, UK

Conference Program Chairs

Kei Eguchi, Fukuoka Institute of Technology, Japan
Zhe Chen, Aalborg University, Denmark
Dimitrios Karamanis, University of Patras, Greece

Publicity Chair

Lihnida Stojanovska-Georgievska, University Ss Cyril and Methodius in Skopje, North
Macedonia

Technical Committees

Contents

Part I
Power Generation and Potential Assessment of Renewable Energy

Optimization of Plug-and-Play PV Systems

David Gögelein and Marianne von Schwerin

Abstract The energy transition, and with it the shift to renewable energy sources, is progressing. However, further measures are needed to accelerate expansion to achieve the European as well as global energy and climate goals. This chapter examines the potential of plug-and-play photovoltaic (PV) systems, which are smaller than traditional PV systems and designed to be installed without the need for professional technical knowledge. As such, this work demonstrates the functionality, applicability and economics of plug-and-play PV systems. Due to the time lag between PV generation and consumption, surpluses often occur. Conventionally, these surpluses are not utilized or remunerated, or only to a limited extent. By integrating energy storage, these surpluses can be accumulated and made available for use at a later time, which can notably increase the self-consumption and thus the economic efficiency of the system. As part of this work, the options for optimizing plug-and-play PV systems with and without energy storage are presented.

Keywords Plug-and-play PV · Optimization · Balcony PV

1 Introduction

The consequences of climate change are noticeable worldwide. Although the share of renewable energies continues to rise, the majority of energy is still being generated from fossil fuels. Furthermore, the European Union imported almost 58% of the energy it consumed because its own production could only meet 42% of its daily needs [1]. To achieve the goals of the energy transition, in particular the reduction of CO_2 emissions along with the reduction of the global mean temperature, further measures must be taken to ensure a sustainable energy supply. The building sector

D. Gögelein (✉) · M. von Schwerin
Ulm University of Applied Sciences, Ulm, Germany
e-mail: David.Goegelein@thu.de; Marianne.vonSchwerin@thu.de

Z. Peng (ed.), *Technological Advancements and Future Directions in Green Energy*, Green Energy and Technology,
https://doi.org/10.1007/978-3-031-83203-1_1

is responsible for 40% of the total primary energy consumption in the world and 24% of the greenhouse gas emissions [2]. When looking at the energy contribution with over 25% of the total energy consumption, the household sector also plays a non-negligible role.

In the past, tenants and apartment owners had little to no opportunity to actively participate in the energy transition. Plug-and-play PV systems, also called balcony PV systems, allow this group not only to actively participate in the transition to clean, renewable energy, but also to reduce their energy costs and dependence on the electricity grid. More drastically, plug-and-play PV is one of very few possibilities for many tenants and homeowners without their own roof space to participate in their own electricity generation from renewable energies [3, 4]. The high number of over two million PV systems in Germany, of which around 64% are small systems with output capacities below 10 kW, shows that these technical possibilities are being used extensively [5]. Photovoltaic is ideally suited to implementing the concept of energy for citizens and strengthening acceptance of the energy transition through participation. According to a representative survey by Lichtblick, solar systems are among the most popular power generation solutions [5]. In the next few years, 1 to 2 GW of decentralized renewable generation capacity could be connected in Germany without expanding the grid, with capabilities reaching 90 GW worldwide [3]. Further studies show similar results, which underline the great potential of plug-and-play PV systems globally [6]. Above all, the hereby achievable cooling effect through avoided greenhouse gas emissions through renewable electricity generation is an important factor in achieving climate neutrality [5].

The following section provides information about conventional plug-and-play PV systems, their design and functionality, as well as their limitations. This is followed by a discussion of options for optimizing plug-and-play PV systems. A simulation and results section analyzes and compares the possibilities for improving existing systems. Finally, the conclusion and discussion section summarizes the main findings of this work, with limitations indicated at the end.

2 State of the Art

Plug-and-play PV systems are, as the name implies, complete sets installed via solar connectors to generate electricity. These systems are intended to make solar energy more accessible and user-friendly, by simplifying the registration and installation process without requiring specialized technical knowledge. Since typically only single PV modules are connected, they may be installed without the need for professional assistance, due to the PV module voltage being in the extra-low voltage region below 120 V DC, as defined in the IEC 61140:2016 including requirements for electrical installations. Therefore, there are no health consequences due to the low voltage in the event of direct and indirect contact. Typically, further safety regulations and standards may be introduced, depending on the location. Such regulations may involve the use of special connectors, electricity meters, and others to comply with safety standards. Furthermore, plug-and-play PV systems are designed to be modular and with it scalable, allowing users to start with a few PV panels and expand their capacity over time. Overall, they offer a cheap alternative to conventional PV systems, making them more applicable for small- and mid-scale applications. At the end of 2023, more than 300,000 plug-and-play PV systems were in operation in Germany [7].

2.1 System Architecture and Functionality

The typical structure of a plug-and-play PV system consists of one or two PV modules that are directly connected to one or two inputs of a micro-inverter via solar connectors. The PV modules generate electricity in the form of direct current (DC) through the photovoltaic effect. The micro-inverter converts the incoming DC into alternating current (AC) in the form of a sinusoidal waveform. The AC output of the micro-inverter is connected to the domestic power supply, e.g., via a Schuko plug. The voltage level and frequency of the AC inverter output are designed to match the conditions of the local power grid, which may differ between Europe and North America, for example. Also, the maximum power which micro-inverters are allowed to supply is limited, e.g., to 600 W or 800 W, again depending on local regulations. During operation, the frequency and voltage of the micro-inverter are synchronized with the power grid and can only supply AC power accordingly, so called grid-tied, which protects users against electric shocks when touching the output terminals, but also means they cannot be used as a stand-alone solution in off-grid applications. The appliances connected to the domestic grid can use this generated power directly for their supply, which means they don't have to be supplied from the external power grid, thus saving energy costs over the course of the year. When little or no PV power is available, the energy required in the household is drawn from the grid as conventional. However, if more PV power is available than required by the household, the surplus energy is fed into the local grid. Depending on the regulation, the amount of energy fed into the grid may or may not be remunerated. The feed-in and consumption data are recorded depending on the electricity meter. The overall components and the functional principle are shown in Fig. 1.

2.2 Limitations and Challenges in the Use of Conventional Plug-and-Play PV Systems

Alongside the mostly positive factors mentioned so far, there are some challenges in the use of plug-and-play PV systems that limit their economic viability and applicability. These include, but are not limited to, feed-in restrictions, no or low remuneration for the energy fed into the grid, regulatory conditions, and the availability of PV energy. These factors are discussed further in the following section.

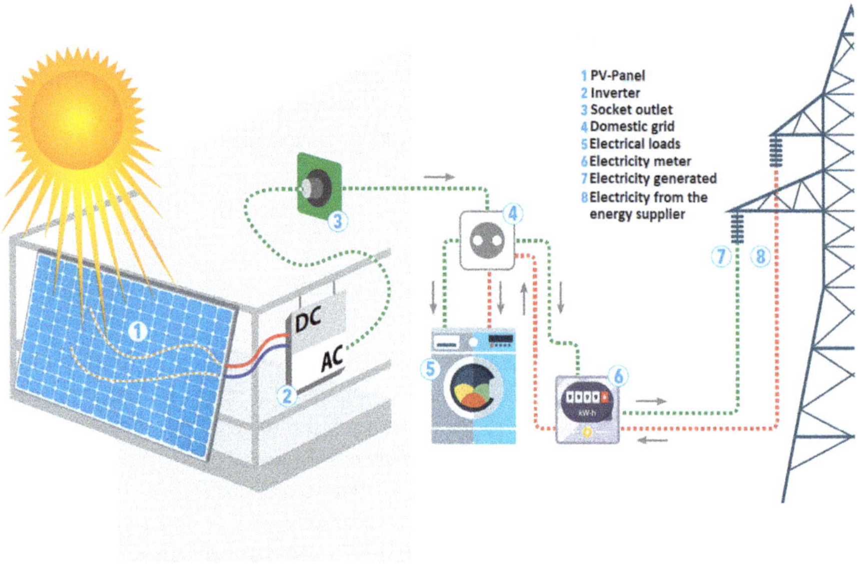

Fig. 1 System components and functional principle of a plug-and-play PV system. (Based on [8])

Feed-in restrictions Plug-and-play PV systems are designed for a maximum permissible feed-in power and throttle their output to this legal maximum accordingly. In most parts of the European Union, this maximum feed-in power is limited to 800 W. In Germany, the permissible limit is still at 600 W, although an increase to 800 W is currently being discussed. Therefore, even if a higher PV power was available, the inverter is configured to not feed in more than this maximum value. Since most households have high-power appliances like electric heaters, cooling systems, and alike, power values exceeding the maximum capability of the micro-inverters cannot be supplied by the micro-inverter alone.

Remuneration Currently, there is generally no or little remuneration for excess electricity, i.e., electricity that is not effectively used to power household appliances and is therefore fed into the grid. Depending on local regulations, it is possible to receive compensation for a part of the normal purchase price, e.g., around 8 cents/kWh in Germany, in agreement with the grid operator and by using dedicated electricity meters [3, 5]. On the other hand, there are even some regulators, such as the Australian Energy Market Commission, who are implementing an export tariff to charge customers for exporting (PV-generated) energy at undesirable periods [4]. To maximize economic efficiency, it is ultimately more lucrative to use the solar power directly in the household.

Regulations Especially in Germany are many regulations for the use of plug-and-play PV systems, which can differ depending on the location or grid operator, causing confusion and therefore acting as a deterrent for many potential users. Some of the regulations include the mandatory use of special electrical equipment like digital electricity meters or specific connectors, e.g., Wieland plugs, which have to be installed by an electrician, ultimately contradicting the basic idea of simple and user-friendly installation. Also, there are above-mentioned feed-in restrictions, whereby in Germany the limit is currently at 600 W while most other European countries are regulated at 800 W. As many countries in the EU have already eased the use of plug-and-play PV systems, Germany is also planning to simplify the use for potential users and therefore make it more attractive. These agreements are summarized in the so-called "Solarpaket 1," which, among other things, provides for the simplification of registration with the grid operator, a clearly defined power limit of 800 W with an allowed maximum installed PV power of 2 kWp, as well as the use of conventional electricity meters and the use of Schuko sockets. The legislation is scheduled to come into effect in 2024 [9]. Nonetheless, local and international regulations restrict the possibilities for the use of plug-and-play PV systems through these boundary conditions and must therefore be considered in their design.

Power availability As the feed-in power of plug-and-play PV systems depends on the amount of solar radiation, they cannot always supply power to cover household needs, especially during nighttime hours. As hinted above, self-consumption is more lucrative, the greater the difference between the purchase costs for electricity and the electricity generation costs of the PV system. For systems without a storage, the self-consumption potential depends on the coincidence between the generation and consumption profile. Depending on the size of the system, households typically achieve 20–40% self-consumption in relation to the electricity generated [5]. Larger systems increase the degree to which the entire electricity demand is covered by PV electricity, but reduce the consumption share [5]. Furthermore, a high concentration of PV systems in a low-voltage grid can lead to electricity production exceeding electricity consumption locally. Transformers then feed power back into the medium-voltage grid. Local clusters of systems in thinly populated areas sometimes require the reinforcement of the grid, transformer stations, or the construction of storage capacities, whereby an even distribution of PV installations across the grid sections reduces the need for grid expansion [5].

2.3 Current State of Knowledge

The current state of knowledge regarding plug-and-play PV systems discussed here focuses on two core areas: firstly, the applicability and cost-effectiveness of plug-and-play PV systems and secondly, technical limitations and optimization options. The most important findings are summarized below.

As for the applicability and cost-effectiveness, several studies found that plug-and-play PV systems have a substantial potential. Mundada et al. [6] found, the total potential US market of over 57 GW, which represents an opportunity for sales for retailers from $14.3 to $71.7 billion depending on the capital cost of plug-and-play solar systems ($0.25 to $1.25/W), which could generate around 108,417,000 MWh/a being four times the electricity generated from US solar in 2015. Similar results were found by [10], investigating the applicability and benefits of plug-and-play PV systems as a framework in the USA, concluding such systems can be installed at a cost of $1.56/W in 2020 [11]. evaluates the practicality of implementing a plug-and-play PV system in Austria versus in Germany. The research concludes that the installation of plug-and-play PV systems in Germany and Austria is a complex issue due to the lack of clear specifications, but that these systems have the potential to play a big role in the ongoing energy transition. The research provided by [12] further shows a high market potential for plug-and-play PV systems, whereby the legal framework remains the largest obstacle for scal ing up tenant electricity. Long-term trends at the landscape level suggest that climate change, rising electricity consumption, and climate protection and technical drivers like smart meters will improve conditions and ultimately increase demand for tenant electricity, resulting in the expansion of rooftop PV generation in urban areas. Khan et al. [13] discusses the barriers in making the plug-and-play system for residential applications and presents an interface concept including a hardware between the grid and the PV panel and a web portal to simplify installation and commissioning, by eliminating physical site-visit and reducing necessary paperwork.

Finally, [4] examines the technical aspects of optimizing the self-consumption of plug-and-play PV systems by using a battery as an energy storage. The research concludes that a remarkable increase in self-consumption, reaching almost 100% can be achieved with a load-controlled battery storage system. Furthermore, the research underlines the long-term benefits by increasing the self-consumption. The following chapters elaborate upon this approach and discuss ways to maximize the cost-effectiveness of plug-and-play PV systems.

3 Plug-and-Play PV System Design Optimization Approaches

One of the main aspects regarding the economic efficiency of a plug-and-play PV system lies in the behavior of the individual using the electricity. Therefore, as mentioned before, it is important to minimize the amount of electricity that is fed into the grid and therefore increase the self-consumption [11]. To achieve this, there are two main approaches: firstly, the alignment of consumption with periods when PV energy is available, and secondly, the temporal compensation between production and consumption by storing the energy like already implied by [4].

3.1 Generation and Demand Alignment

There are two ways to influence the synchronization of generation and consumption without a storage unit: firstly, adjusting the timing of electricity generation by using a dedicated PV installation and secondly, controlling the timing of energy consumption.

Influencing the time of power generation Depending on the installation parameters of PV modules, e.g., orientation and inclination, the power generation curve shifts. Thereby, installations with east-orientation have higher output during the morning hours, south- and north-orientation have their peak during noon, while west-oriented modules have higher PV generation during the evening hours. Even though the installation parameters of PV modules are one way to influence the power generation using PV, the possible installation variants are limited by the circumstances at the installation site. Users with flexible installation options can increase their self-consumption rate by aligning the power generation curve with their consumption behavior, e.g., west-oriented PV modules for users who typically need most of their energy during evening hours.

Influencing the time of power consumption As an alternative to adjusting the power generation curve of the PV modules, the user has the option of shifting the demand to a time of availability. This includes topics such as switchable loads, e.g., via smart Wi-Fi sockets, the manual use of household appliances at peak production times, e.g., cooking or washing machine, as well as automated controls. This process is also known as load management [5]. One way to implement such a process is via smart meters, for instance, by measuring what is fed into and drawn from the grid. Whenever a surplus is measured, consumers are switched on as required. You can, for example, set your own threshold values for the power at which consumers are switched on. If there is no longer a surplus, the loads are simply switched off again automatically. As this approach requires knowledge of generation, consumption and general technical characteristics, and plug-and-play PV systems are especially designed for non-technical users, this should not be considered the norm, but it offers users a good opportunity to increase their economic efficiency of their plug-and-play PV system.

3.2 Energy Storage

An alternative to reducing the imbalance between energy demand and energy production is by storing energy. Conventionally, this is done by either using electrochemical storage like batteries or by converting and storing the energy thermally, e.g., in hot water tanks. Storage system installations are done to increase the self-consumption rate, since the price for drawing electricity from the grid is notably higher than the remuneration for electricity fed into the grid [4]. Currently, plug-in PV systems are rarely combined with a form of energy storage, approximately 7% in 2023 according to [4]. Due to the fluctuating nature of solar energy generation, the self-consumption rate without an energy storage system varies between 30% and 37% for an average European household [4], whereby higher rates can be achieved when the PV system is well-matched with the base load of the household [4]. By integrating an energy storage, the initial costs are typically higher, but the increase in self-consumption rate and reduction of electricity drawn from the grid, will lead to higher economic efficiency eventually. One crucial task in this regard is designing the size of the energy storage system appropriately, as an energy storage system that is too large is often underutilized and unnecessarily increases costs, while one that is too small leads to a capping of solar generation too often [4]. Further, conventionally mostly batteries are considered as energy storage in plug-and-play PV systems as the electric energy they provide is of high value, yet thermal storage can have high capacities while keeping the storage costs comparatively lower.

The solutions proposed so far require active user involvement and technical knowledge. Plug-and-play PV systems are meant for non-technical end users, which is why the user should not have to adapt to the system, but rather the other way around. Therefore, the authors of this chapter suggest the idea of introducing an intelligent modular plug-and-play PV system utilizing a combined thermal and electrical energy storage with energy management to achieve a highly economical and easy to use energy system.

4 Simulation and Results

To test the utilization potential and the economic efficiency of conventional plug-and-play PV systems with and without storage, simulations were conducted. For this purpose, PV generation data from PVGIS was used, for which the location Ulm was looked at with a south-facing system of 1 kWp at an inclination of 30° [14]. The most recent available data period was chosen, commencing on January 1, 2019, and concluding on December 31, 2020, a total of 2 years.

Though this does not represent each household due to the wide variety of users and their electricity usage habits, it can serve as a reference. In the simulation, PV systems with an installed peak power of 400 to 2000 Wp were examined, firstly without a storage and secondly with batteries ranging from 1 to 3 kWh storage capacity. Thus, the resulting PV surplus was examined based on the annual amount of available solar energy and user demand, whereby each possible configuration was assessed, as illustrated in Fig. 2. The simulation is used to see how much of the PV electricity generated can be used directly in the household and how much surplus is generated and whether it can be stored or fed into the grid using a micro-inverter with a maximum feed-in power of 800 W. In the simulation, the coverage of the electrical demand was prioritized, and excess energy was charged into a battery with a maximum C-rate of 0.5, defining the allowed maximum charging and discharging current in dependence of the battery capacity. During the simulation, energy stored in the battery is withdrawn when less PV is available than demand. If the battery is full or the maximum charging power is exceeded, remaining PV power is fed into the micro-inverter for partial compensation until the maximum limit of 800 W is reached. Any further available power from the PV modules cannot be utilized. The resulting energy savings were then compared with the investment costs and a payback period was calculated, based on the assumptions shown in Table 1. Depending on the individual case, these numbers can vary widely, which is why the

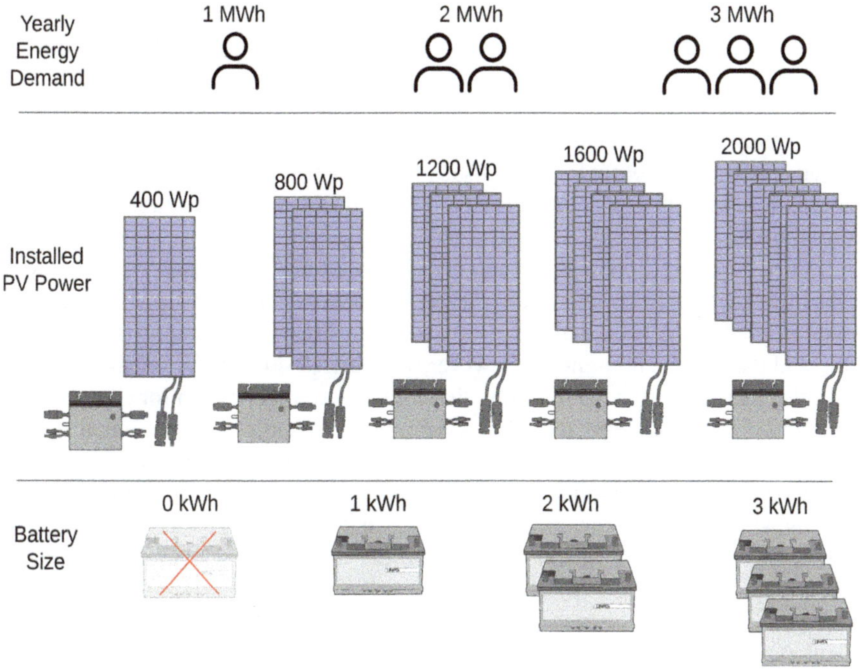

Fig. 2 Simulation variables and evaluated system configurations

Table 1 Underlying assumptions used in the simulations

Description	Value
Micro-inverter cost	150 €
PV cost per kWp	150 €
Fixed PV installation cost	200 €
Cost per kWh of electricity	0.30 €
Remuneration per kWh	0.082 €
Cost per kWh battery storage	500 €
Maximum C-rate	0.5

energy cost savings per year are also shown in the following so that they can be adapted to other scenarios.

The simulation results depend on the installed PV power, the annual energy demand, and the size of the battery storage system. Remunerable and non-remunerable surplus is calculated as a percentage of the annual PV energy, as well as the approximate payback period and the annual energy cost savings. The annual PV energy for the given location and installation parameters is roughly 1100 kWh/a per installed peak power.

The selected installation parameters with south-orientation and an inclination of 30° to the horizontal are commonly used in Germany to maximize the annual electricity yield [5]. By comparison, PV modules installed otherwise typically produce a lower yield, e.g., on a west-facing roof with a 40° tilt around 24% less yield, leading to higher electricity generation cost, which extends the payback period [5]. Due to the high efficiency of micro-inverters of around 98%, these as well as other system losses were not incorporated in the simulation [5]. The simulation results in Table 2 show that with increasing annual energy demand, more energy is consumed directly, and consequently less surplus is generated. Considerable non-remunerable surpluses arise with higher installed PV power due to the power limitation of the micro-inverter.

Tables 3, 4, and 5 summarize the simulation results with battery storage with capacities ranging from 1 to 3 kWh. Here, an increase in PV installation capacity leads to a reduction in annual energy costs due to simply more energy being available and used in the household. However, with increased installation size, the surplus also increases. It is worth noting that a battery storage system for conventional plug-and-play PV systems with 400 kWp is not profitable with an annual energy demand of more than 2 MWh. As reference, an average household electricity consumption for electrical appliances, lighting, hot water (hygiene purposes), and space heating was 1.6 MWh per household member in 2018. Average values for 1-person households are slightly higher per person, for multi-person households notably lower [5].

Figure 3 shows the surplus in relation to the installed PV power, the annual electrical energy demand, and various battery sizes. The surplus shown here includes excess energy that cannot be used or stored directly, and is made up of the partially remunerated and non-remunerated share. This surplus could be used more efficiently, e.g., by introducing a thermal energy storage, like a water heater or a heat pump. Energy stored thermally in latent heat storage is described in formular 1,

Table 2 Simulation results without storage

Energy demand (MWh)	Installed PV power (kWp)	Surplus remun. (%)	Surplus unrem. (%)	Payback period (a)	Energy savings (€/a)
	0.4	26	0	4.15	99
	0.8	55	0	3.90	121
1	1.2	66	2	4.03	132
	1.6	63	11	4.26	139
	2.0	58	21	4.54	143
	0.4	1	0	3.09	132
	0.8	26	0	2.38	198
2	1.2	43	2	2.36	224
	1.6	44	11	2.45	241
	2.0	42	21	2.56	253
	0.4	0	0	3.06	139
	0.8	1	0	1.91	247
3	1.2	25	2	1.79	297
	1.6	28	11	1.81	326
	2.0	28	21	1.88	346

Table 3 Simulation results with 1 kWh battery storage

Energy demand (MWh)	Installed PV power (kWp)	Energy stored (%)	Energy direct-use (%)	Surplus remun. (%)	Surplus unrem. (%)	Payback period (a)	Energy savings (€/a)
	0.4	26	74	0	0	6.80	134
	0.8	27	45	28	0	4.51	215
1	1.2	21	33	46	0	3.85	267
	1.6	17	26	48	9	3.63	300
	2.0	15	21	45	19	3.57	322
	0.4	1	99	0	0	6.80	134
	0.8	19	74	7	0	3.81	255
2	1.2	18	56	25	1	3.16	326
	1.6	16	45	30	9	2.95	369
	2.0	13	38	31	18	2.88	399
	0.4	0	100	0	0	6.80	134
	0.8	8	92	0	0	3.63	267
3	1.2	15	73	11	1	2.80	367
	1.6	14	61	17	8	2.56	425
	2.0	13	51	19	17	2.48	464

Table 4 Simulation results with 2 kWh battery storage

Energy demand (MWh)	Installed PV power (kWp)	Energy stored (%)	Energy direct-use (%)	Surplus remun. (%)	Surplus unrem. (%)	Payback period (a)	Energy savings (€/a)
	0.4	26	74	0	0	10.5	134
	0.8	37	44	19	0	6.33	232
1	1.2	29	33	38	0	5.29	289
	1.6	23	26	42	9	4.9	325
	2.0	19	21	41	19	4.74	348
	0.4	1	99	0	0	10.5	134
	0.8	26	74	0	0	5.5	268
2	1.2	31	56	13	0	4.2	362
	1.6	27	45	21	7	3.8	419
	2.0	24	38	23	15	3.6	458
	0.4	0	100	0	0	10.5	134
	0.8	8	92	0	0	5.5	268
3	1.2	23	74	3	0	3.9	392
	1.6	23	61	11	5	3.4	468
	2.0	22	51	14	13	3.2	518

which depends on the mass m, the heat capacity c_p, and the temperature difference. For water in the temperature range from 0 to 100 °C, the heat capacity is $4.19\ \frac{\text{kJ}}{(\text{kg} \cdot \text{K})}$ [15].

Table 5 Simulation results with 3 kWh battery storage

Energy demand (MWh)	Installed PV power (kWp)	Energy stored (%)	Energy direct-use (%)	Surplus remun. (%)	Surplus unrem. (%)	Payback period (a)	Energy savings (€/a)
	0.4	26	74	0	0	14.27	134
	0.8	38	45	17	0	8.32	237
1	1.2	31	33	36	0	6.86	296
	1.6	25	25	41	9	6.31	331
	2.0	21	21	40	18	6.07	354
	0.4	1	99	0	0	14.27	134
	0.8	26	74	0	0	7.36	268
2	1.2	37	56	7	0	5.31	382
	1.6	34	45	16	5	4.66	449
	2.0	31	38	19	12	4.35	494
	0.4	0	100	0	0	14.27	134
	0.8	8	92	0	0	7.36	268
3	1.2	26	74	0	0	5.07	401
	1.6	31	61	6	2	4.17	501
	2.0	29	52	10	9	3.82	563

$$Q = m \cdot c_p \cdot (T_1 - T_2) \tag{1}$$

Consequently, 1 kWh of energy is enough to heat approximately 20 l of water from an initial temperature of 20 to 60 °C with a water heater using a PV heating rod. Looking at the surplus of a system size with a PV installation of 1.2 kWp, an annual energy demand of 2 MWh and a battery storage size of 1 kWh, around 7 tons of hot water could potentially be produced for domestic use annually. Depending on the system configuration and the amount of surplus energy generated as well as the demand, it is practical and economical to use a combined electrical and thermal storage system. Such a system would require a user-orientated energy management system to distribute the available energy as economically as possible. Hereby, the hot water storage serves as a cost-effective way of storing large amounts of energy and thus using the resulting surplus. Since the available energy depends on the time of year, a grid supplementary heating system should be used to meet the demand, as the energy would not be sufficient throughout the whole year.

5 Conclusion and Discussion

This work shows the applicability and economic efficiency of plug-and-play PV systems. Conventionally, due to the time lag between PV generation and consumption, surpluses occur frequently, which are typically not remunerated or compensated only marginally. By integrating energy storage, these surpluses can be

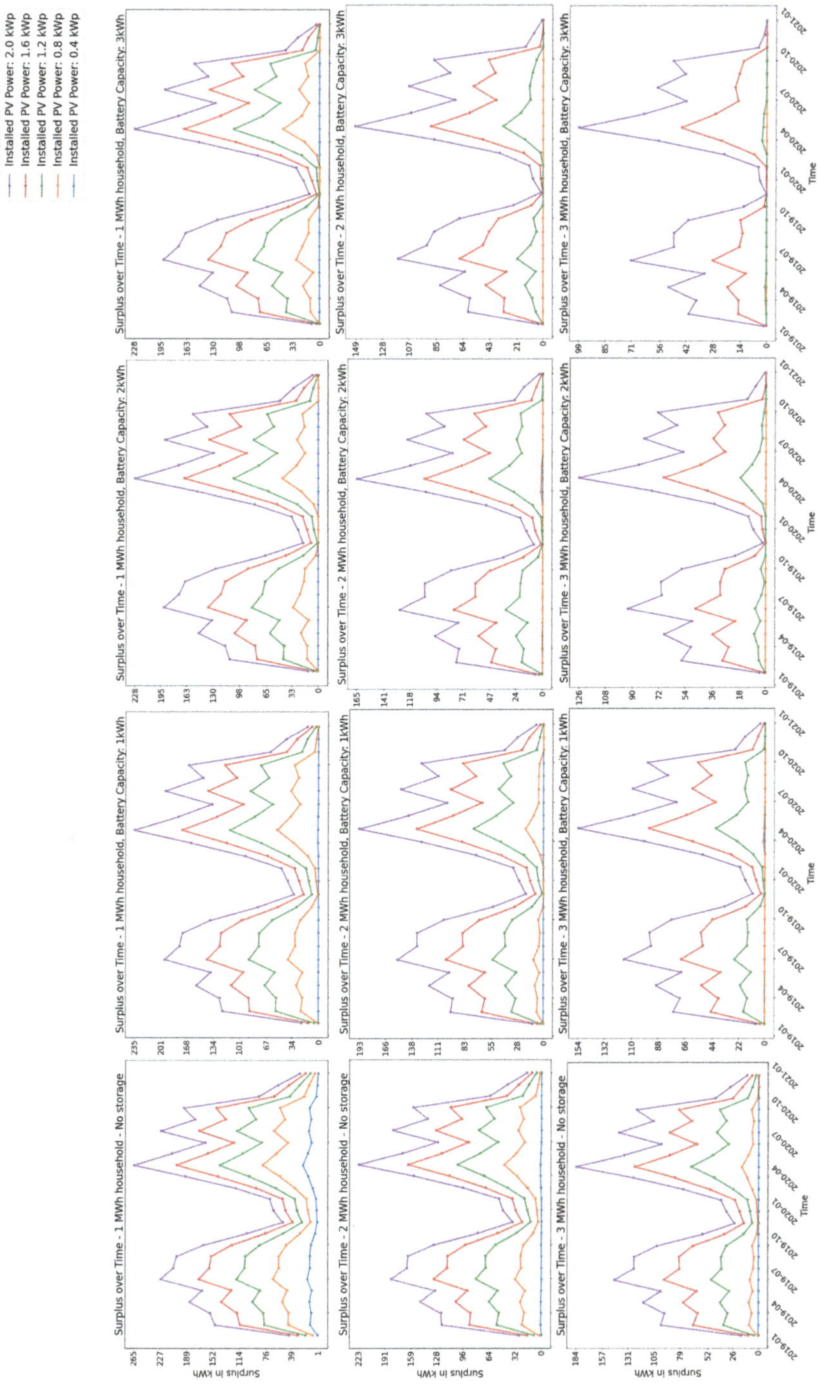

Fig. 3 Total energy surplus depending on the installed PV power, battery size, and energy demand

accumulated and made available for use at a later time, which can increase self-consumption and thus the economic efficiency of the system. As part of this work, the options for optimizing plug-and-play PV Systems with and without energy storage were presented. Hereby, a simulation based on PV generation data from PVGIS and the standard load profile for households H0 was conducted [16]. The simulation shows that especially in the summer months, a substantial amount of non- or only partially compensated surplus is generated with and without battery storage, depending on the installed PV power, the annual user demand, and the storage size. Furthermore, the study found that the use of batteries can be economical in the long term while having a higher initial cost. Also, the additional integration of thermal energy storage, such as a hot water tank, can further increase the economic efficiency of the system by storing surpluses that cannot be utilized electrically, due to the maximum power feed-in by micro-inverters and limited battery capacity. To satisfy electricity demand and efficiently store energy, an energy management system is essential to optimally distribute energy following user behavior and demand. Overall, plug-and-play PV systems have great potential to cover a considerable share of the energy demand of homes with little space for PV installations at low system costs, which further helps to support the energy transition.

6 Limitations

The electrical consumption data used in this research is based on aggregated consumption patterns, which may result in deviations in time-based electrical demand depending on the user's behavior. Furthermore, both the PV generation and load profile data used are based on conditions in Germany. The approach for optimizing plug-and-play PV systems and the assessment of economic efficiency can be transferred to other countries and conditions. However, local data should be used for such an evaluation.

References

1. Dimara, A., Sougles, C., Athanasiou, S. et al. Holistic plug-n-play autonomous solar system integration: a real-life small-scale demonstration—a practical approach. Electr Eng 105, 2715–2733 (2023). https://doi.org/10.1007/s00202-023-01830-6.
2. C. Vassiliades, G. Barone, A. Buonomano, C. Forzano, G.F. Giuzio, A. Palombo, Assessment of an innovative plug and play PV/T system integrated in a prefabricated house unit: Active and passive behaviour and life cycle cost analysis, Renewable Energy, Volume 186, 2022, Pages 845–863, ISSN 0960-1481.
3. Deutsche Gesellschaft für Sonnenenergien - PVplug.de, https://www.pvplug.de/. Last accessed 15 Jan 2024.
4. U. Spaeth, A. Popp, H. Fechtner, A. Cichon and B. Schmuelling, "Self-consumption Optimization of Balcony Solar Systems Using a Load-controlled Battery Storage," 2023 10th International Conference on Power and Energy Systems Engineering (CPESE), Nagoya, Japan, 2023, pp. 378–382, https://doi.org/10.1109/CPESE59653.2023.10303065.

5. Harry Wirth, Aktuelle Fakten zur Photovoltaik in Deutschland, Fraunhofer ISE, Download von www.pv-fakten.de, Fassung vom 20. Dezember 2023.
6. Aishwarya S. Mundada, Emily W. Prehoda, Joshua M. Pearce, U.S. market for solar photovoltaic plug-and-play systems, Renewable Energy, Volume 103, 2017, Pages 255–264, ISSN 0960-1481.
7. Janson Matthias, Statista: Mini-Solaranlagen, https://de.statista.com/infografik/30601/kumulierte-anzahl-der-in-betrieb-befindlichen-balkon-solaranlagen-in-deutschland/, 29.08.2023. Last accessed 15 Jan 2024.
8. IMTEST- Graphik grüner Strom vom Balkon, https://www.imtest.de/wp-content/uploads/sites/7/2023/04/graphik-gruener-strom-vom-balkon.jpg. Last accessed 21 Feb 2024.
9. Bundesregierung - Solarpaket I, https://www.bundesregierung.de/breg-de/aktuelles/solarpaket-photovoltaik-balkonkraftwerke-2213726. Last accessed 24 Jan 2024.
10. Fraunhofer USA - Plug and Play PV, https://www.cmi.fraunhofer.org/pnp. Last accessed 15 Jan 2024.
11. Breit, T. P. (2023). Balcony PV power plants in Austria: Public support for adoption and differences in implementation between Austria and Germany [Master Thesis, Technische Universit¨at Wien; Diplomatische Akademie Wien]. reposiTUm. https://doi.org/10.34726/hss.2023.113066.
12. Moser, R.; Xia-Bauer, C.; Thema, J.; Vondung, F. Solar Prosumers in the German Energy Transition: A Multi-Level Perspective Analysis of the German 'Mieterstrom' Model. Energies 2021, 14, 1188. https://doi.org/10.3390/en14041188.
13. M. T. A. Khan, I. Husain and D. Lubkeman, "Power electronic components and system installation for plug-and-play residential solar PV," 2014 IEEE Energy Conversion Congress and Exposition (ECCE), Pittsburgh, PA, USA, 2014, pp. 3272–3278, https://doi.org/10.1109/ECCE.2014.6953845.
14. European Commission - Photovoltaic Geographical Information System (PVGIS), https://joint-research-centre.ec.europa.eu/photovoltaic-geographical-information-system-pvgis_en. Last accessed 19 Feb 2024.
15. Sterner, Michael and Stadler, Ingo, Energiespeicher-Bedarf, Technologien, Integration. Springer, 2017, ISBN: 978-3-642-37379-4.
16. BDEW - Standardlastprofile Strom, https://www.bdew.de/energie/standardlastprofile-strom/. Last accessed 19 Feb 2024.

Design of Load Management Methods Using Distributed Storage and Their Impact on The Aging Level of Distribution Transformers

Ivan David Endara, Diego Maldonado Macas, Miguel Parra Valverde, and Jorge Montealegre García

Abstract The objective of this study is to analyze the influence of various parameters on the expected lifespan of transformers, assess the economic viability for utilities and end-users to implement photovoltaic generation systems, and measure the extension of transformer lifetime. For power distribution transformers, understanding the working temperature and humidity is crucial. Abnormal values of these parameters result in electrical and mechanical stress, ultimately affecting the transformer's lifespan.

As an alternative, distributed solar generation systems with storage in batteries were explored. Different scenarios involving solar panels and inverters were analyzed to determine the optimal solution for load relief, reduction in operating temperature, and improvement in the utilization factor, consequently enhancing the overall lifespan.

Keywords Transformer aging · Utilization factor · Residential solar generation system · Hot-spot temperature

1 Introduction

The functioning of transformers, from the onset of their operation, is governed by parameters or variables. Any deviation from normal levels of operation can lead to malfunctions, prompting operators to make timely decisions. This document studies these variables and their variations over time, forming a matrix. The matrix is linked to the aging percentage of electrical equipment, aiding in determining the expected lifespan.

I. D. Endara (✉) · D. M. Macas · M. P. Valverde · J. M. García
Escuela Superior Politécnica del Litoral, Guayaquil, Ecuador
e-mail: iendara@espol.edu.ec; diefmald@espol.edu.ec; miadparr@espol.edu.ec; montealegre@espol.edu.ec

Z. Peng (ed.), *Technological Advancements and Future Directions in Green Energy*, Green Energy and Technology,
https://doi.org/10.1007/978-3-031-83203-1_2

Another critical parameter under consideration is the load curve, representing user behavior in electricity consumption. To manage distributed storage load effectively, a method is proposed to decrease and smooth load curve peaks observed by substation distribution transformers. Consequently, an analysis is conducted to identify the optimal timing for utilizing photovoltaic energy and its impact on transformers.

2 Methodology

The project focuses on studying power transformers by analyzing internal parameters such as temperature variation in the hottest point of the winding and the percentage of humidity inside the equipment. A matrix is designed to depict these variables over time, providing operators with insights into their behavior and facilitating preventive maintenance planning.

To do so, two different approaches are used to model the operational status of transformers.

2.1 *Hot-Spot Analysis*

Using Std. ANSI/IEEE C57.92–1981, it is possible to calculate the difference at the hot-spot and the oil temperature at the top of the tank based on the load variations, presented on the Eq. 1:

$$\Delta\theta_{\mathrm{H}} = \Delta\theta_{\mathrm{HS,R}} \times K^{2m} \tag{1}$$

where,

θ_{H}: hot-spot temperature
$\Delta\theta_{\mathrm{HS,R}}$: difference between the hot-spot temperature and oil temperature on the top at nominal load.

$$\Delta\theta_{\mathrm{HS,R}} = \Delta\theta_{\mathrm{hotspot\ winding}} - \theta_{\mathrm{Top\,oil}}$$

K: specific load in p.u.

Then it is possible to calculate the maximum hot spot temperature for windings for each load variation over time, as shown on Eq. 2:

$$\theta_{\mathrm{H}} = \Delta\theta_{\mathrm{TO}} + \Delta\theta_{\mathrm{H}} + \theta_{A} \tag{2}$$

where,

$\Delta\theta_{\mathrm{TO}}$: difference between top-oil temperature and environment temperature.

θ_A: environment temperature.

Finally, it is possible to evaluate the temperature variation over time which significantly impacts transformer lifespan, as a considerable increase results in a high aging factor. Equation 3 illustrates the relationship between the aging factor and temperature.

$$F_{AA} = e^{\left[\frac{15000}{383} - \frac{15000}{\theta_H + 273}\right]} \tag{3}$$

2.2 *Humidity Analysis*

The percentage of humidity inside the transformer is another parameter under study. Water particles, located predominantly in the insulating paper and to a lesser extent in the mineral oil, influence dielectric resistance [3]. The equation below depicts moisture behavior in the dielectric paper [2], determining its aging factor (Table 2).

$$\mathbf{FAA} = \left(\frac{\mathbf{1}}{\mathbf{DP_{END}}} - \frac{\mathbf{1}}{\mathbf{DP_{START}}}\right) * A * e^{\left(-\frac{13350}{T+273}\right)} \tag{4}$$

where:

DPend: Final polymerization degree of the paper (200). This degree of paper polymerization means the end of the dielectric paper's useful life.
DPstart: Initial polymerization degree of the paper (1000–900). With this level of polymerization, the transformers start to work.
T: Dielectric paper temperature.
A: Factor that depends on the degree of humidity of the dielectric paper (Table 1).

"The purpose of cellulose in living tissues is to provide a rigid 'cell' to the cell so that it does not expand indefinitely due to the positive osmotic pressure it possesses. This is achieved through a high degree of polymerization. The polymerization grade (GP) values found are in the order of several thousand glucose units. This is why cellulose is called 'high polymer.'" [5]

Table 1 Percentage of moisture on paper

Moisture (%)	A
Dry	1.07×10^8
1.0%	3.50×10^8
1.5%	6.27×10^8
2.0%	7.80×10^8
3.0%	2.03×10^8
4.0%	3.65×10^9

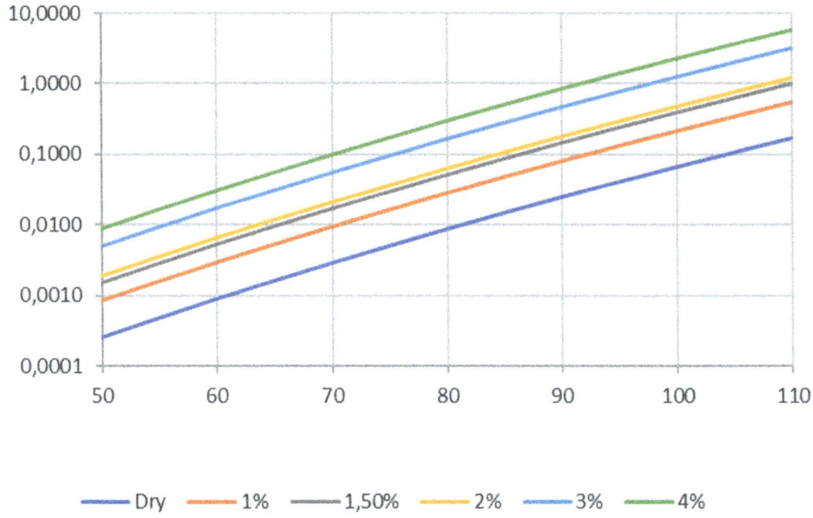

Fig. 1 Aging factor due to humidity [4]

Aging factors increase as the oil temperature rises. It's important to note that this doesn't occur due to periodic maintenance of distribution transformers. Therefore, the aging factor for moisture incursion into the interior, in the worst-case scenario, is considered as 1% (Fig. 1).

The temperature parameters of the hottest point of the winding, due to loadability and the percentage of humidity, influence the useful life of the equipment, decreasing it over time. For this study, the effect of maintenance is not considered, as it introduces modifications in the transformer study during each maintenance cycle. Table 2 interrelates these variables to form a comprehensive matrix.

2.3 Aging Evaluation

As observed in the aging matrix on Table 2, the percentage of moisture does not determine any large-scale variation compared to the factor due to loadability.

In the other hand, it is important to define the annual demand increase in order to correlate the aging matrix into the continuous evaluation of real-time loadability for transformers. As a result, it is possible to define the expected lifetime for the equipment at current and future behavior.

Finally, as distributed generation with load-shifting tanks to the storage system is applied, a new lifespan for power transformers can be obtained.

Table 2 Aging factors matrix

		Oil moisture					
		Dry	1%	1.05%	2%	3%	4%
Hot-spot temperature at a percentage of nominal load	40	0.0002	0.0002	0.0002	0.0002	0.0002	0.0002
	45	0.0003	0.0003	0.0003	0.0003	0.0003	0.0003
	50	0.0007	0.0007	0.0007	0.0007	0.0007	0.0007
	55	0.0014	0.0014	0.0014	0.0014	0.0014	0.0014
	60	0.0028	0.0028	0.0028	0.0028	0.0028	0.0028
	65	0.0054	0.0054	0.0054	0.0054	0.0054	0.0054
	70	0.0104	0.0104	0.0104	0.0104	0.0104	0.0104
	75	0.0195	0.0195	0.0195	0.0195	0.0195	0.0195
	80	0.0358	0.0358	0.0358	0.0358	0.0358	0.0358
	85	0.0649	0.0649	0.0649	0.0649	0.0649	0.0650
	90	0.1156	0.1156	0.1156	0.1156	0.1157	0.1157
	95	0.2026	0.2026	0.2026	0.2026	0.2027	0.2028
	100	0.3499	0.3499	0.3499	0.3500	0.3500	0.3502
	105	0.5957	0.5957	0.5958	0.5958	0.5959	0.5961
	110	1.0000	1.0001	1.0001	1.0001	1.0004	1.0007
	115	1.6565	1.6566	1.6567	1.6567	1.6571	1.6575
	120	2.7089	2.7091	2.7092	2.7092	2.7098	2.7105

2.4 *Distributed PV Generation*

Based on domestic demand, PV generation will be applied at the load in small quantities, in order to relief the transformer utilization. The reason why it is important to define correctly the amount of domestic demand connected to each transformer to be analyzed.

On the other hand, also batteries will be considered as a storage system. The idea to do so is to switch the hours when the PV produced energy will be delivered into the grid, meaning that the load consumption will decrease during these time periods when the transformer is highly loaded.

It will be considered the implementation of different cases of PV generation for domestic loads, using 255 W solar panels. Case 1: 6 panels +0.6 kW inverter, Case 2: 6 panels +1 kW, Case 3: 9 panels +0.75 kW, Case 4: 9 panels +1.5 kW.

3 Case Study

The study area encompasses the Samborondon system, starting from the subtransmission line connecting the distribution system with the Ecuadorian National Interconnected System (SNI).

Figure 2 illustrates the area from the Ciudad Celeste substation to the La Puntilla substation. Notably, all analyzed transformers are delta-star connection transformers.

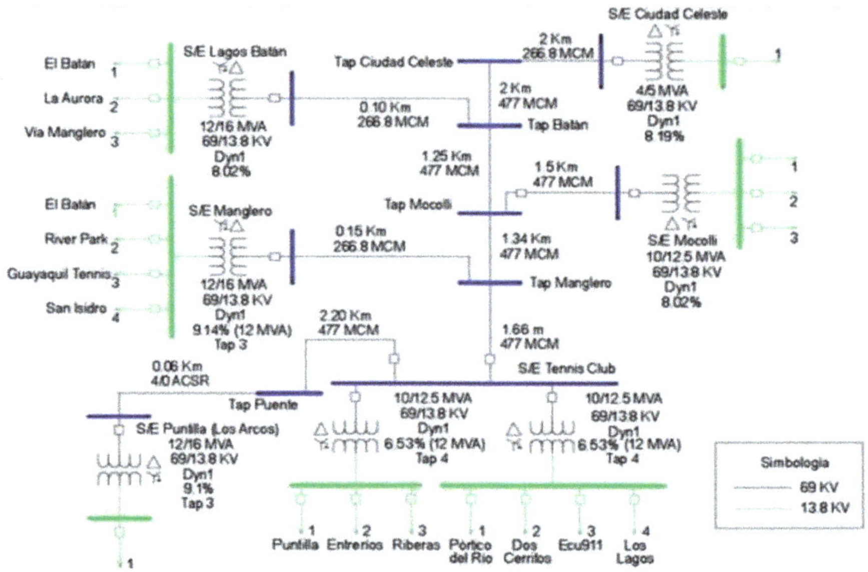

Fig. 2 One-line diagram

Table 3 presents the technical data of the analyzed transformers, each having different operating parameters, aiding in distinguishing temperature behavior due to loadability.

Real historical data on the demand curve, supplied by CNEL. Guayas-Los Rios, were obtained (Fig. 3). The graphs represent the daily load curves on averaged every 10 minutes for April and show that the Manglero substation has the highest demand, while Mocoli has the lowest loadability, making them prime candidates for analysis.

The graphs represent the daily load curves on average every 10 minutes for April and show that the Manglero substation has the highest demand, while Mocoli has the lowest loadability, making them prime candidates for analysis.

The highest temperature of the winding in the transformer signifies the point of greatest thermal and mechanical stress. The internal temperature's non-linear behavior, dependent on loadability, is determined by referencing IEEE std., C57.91–204 [1].

The temperature variation graph for the Manglero substation transformer (Fig. 4) illustrates a maximum temperature of 74 °C due to a peak demand of 0.70 (in p.u). In contrast, the Mocoli substation transformer (Fig. 5), with a lower load, exhibits a lower temperature behavior, reaching approximately 45 °C at a load of 0.16 (in p.u).

Thanks to the growth factor of previous years, the national electrification plan of Ecuador can determine the current status of the transformers by projecting demand into future years with the growth factor obtained in the "Cantonal Plan for Development and Territorial Rearrangement 2015–2019." This helps determine the final year of life for the two equipments based on loadability, a variable that significantly affects the transformer in its operation. As seen in the following graphs, the

Table 3 Technical data of transformers

Substation	Los Arcos	Tenis Club A	Tenis Club B	Manglero	Mocoli	Lagos del Batan	Ciudad Celeste
Year	2015	1996	2014	1998	2009	2010	2009
Brand	TEKSA Transformer	Fuerza Transformer	TEKSA Transformer	Mitsubishi Electric Corporation	Trafo E. E.	Trafo E. E	GEM
Capacity (MVA)	18/24	10/12,5	16/20/24	12/16	10/12,5	12/16	12/16
Cooling system	ONAN/ ONAF	ONAN/ ONAF	ONAN/ ONAF/ OFAF	ONAN/ ONAF	ONAN/ ONAF	ONAN/ ONAF	ONAN/ ONAF
Core and winding Weight (Kg)	17,000	10,400	14,550	14,300	8300	10,010	16,572
Tank Weight (Kg)	12,000	5100	10,500	7500	6600	7040	8328
Oil Volume (Lts)	9227.22	6113.03	9457.9	8300	4152.25	5478.66	9457.9
Nominal Load Losses ONAN (W)	53,747	69,306	56,125	75,995	69,306	75,995	75,995

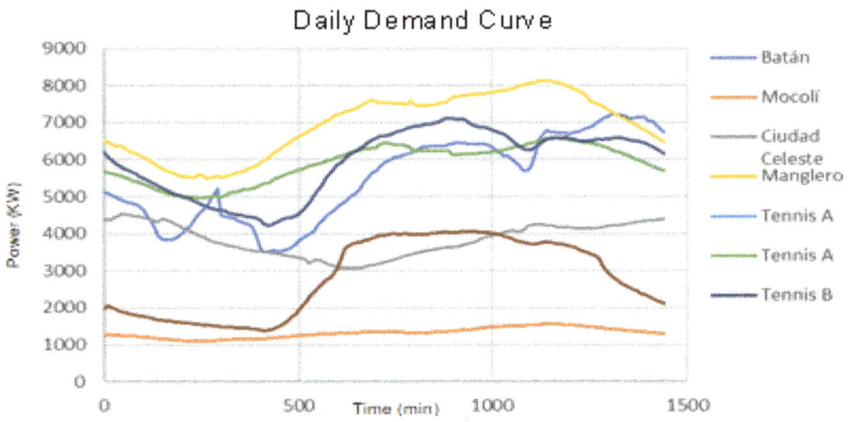

Fig. 3 Demand curve of each substation

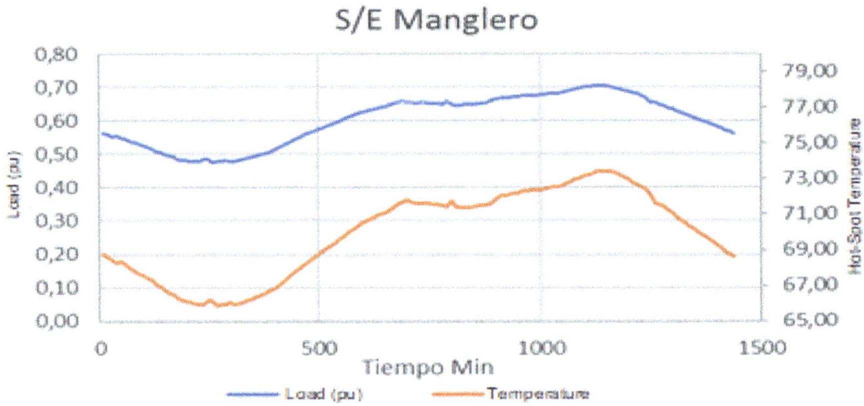

Fig. 4 Temperature vs. loadability of the transformer in the Manglero S/E

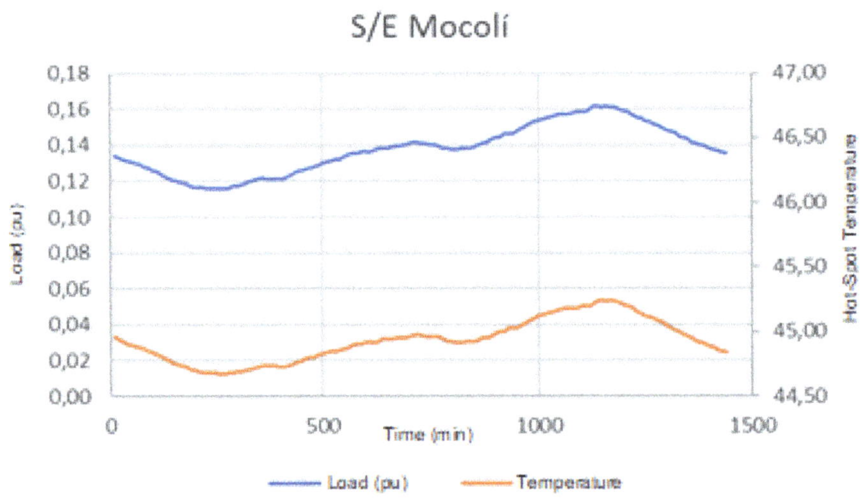

Fig. 5 Chargeability curve at transformer temperature in Mocoli S/E

increases in demand over the years lead the Manglero substation transformer to 100% aging in 2027, while the Mocoli substation transformer reaches this point in 2056. The substantial difference between the two is due to loadability; temperatures of the hottest point for values less than 0.2 p.u. are negligible, while exceeding 0.8 p.u. brings temperatures close to 110 °C, a point where aging factors suddenly increase.

To obtain the variation concerning time, demand projections were made for years before and after the current one for both the Manglero and Mocoli substation

Table 4 Demand forecast (Manglero) [6, 7]

Year	Annual demand increase (%)
2005–2011	4.9
2011-forward	6.9

Table 5 Demand forecast (Mocoli) [6, 7]

Year	Annual demand increase (%)
2009–2011	4.9
2011-forward	5.6

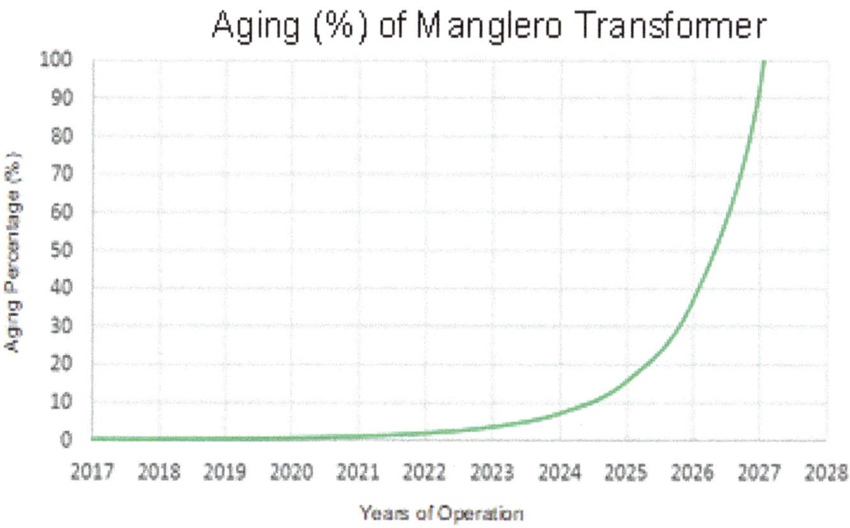

Fig. 6 Aging curve Manglero transformer

transformers. It is worth mentioning that the Manglero transformer commenced operations in 2005, while the Mocoli transformer started operating in 2009 (Tables 4 and 5 and Figs. 6 and 7).

Our aim is to reduce this aging percentage and shift the aging curve to the right, increasing the transformer's lifetime. This is achieved by using solar photovoltaic generation systems with distributed storage (batteries), serving to reduce peaks in demand. The Manglero substation transformer supplies approximately 1300 residential subscribers, while that of Mocoli supplies 400 subscribers. All subscribers were considered to use the same generation system with 6 or 9 solar panels and inverters ranging from 0.6 to 1.5 KW.

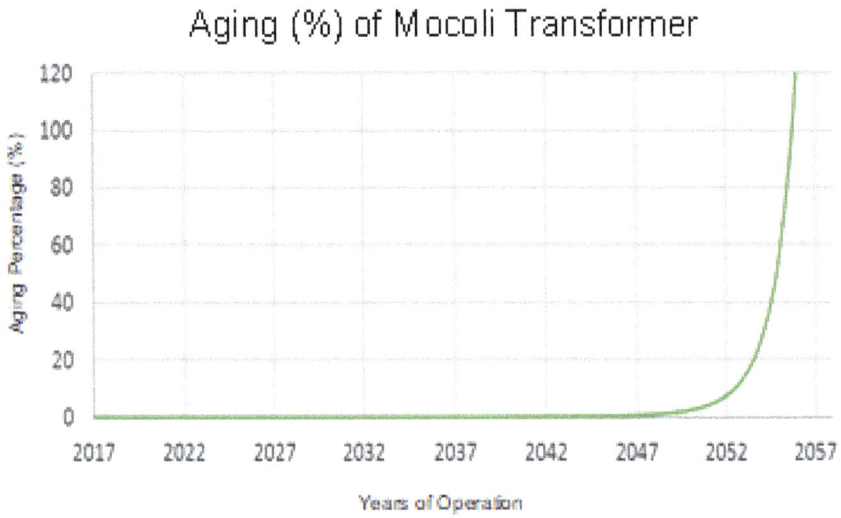

Fig. 7 Aging curve Mocoli transformer

Table 6 Utilization factor reduction

Case	# Panels	Inverter	Utilization factor
Case 1	6 panels	600 W	0.61
Case 2	6 panels	1000 W	0.64
Case 3	**9 panels**	**750 W**	**0.59**
Case 4	9 panels	1500 W	0.62
No panels			0.68

4 Results

4.1 Manglero Option

To size the generation system per resident, an inverter analysis was conducted between 0.6 KW and 1 KW [9] for the set of six 255 Wp [8] solar panels. For the design of 9255 Wp solar panels per home, power delivery to the user varied between 0.75 KW and 1.5 KW inverters. The following figure illustrates that implementing a generation system with 9 solar panels and an inverter of 0.75 KW per subscriber will result in an increase of approximately 3 months with 22 days in the life of the equipment. This represents the best response among the options mentioned. It's worth noting that the distributed storage system operates at night, given the observed peak during that time (Table 6 and Figs. 8 and 9).

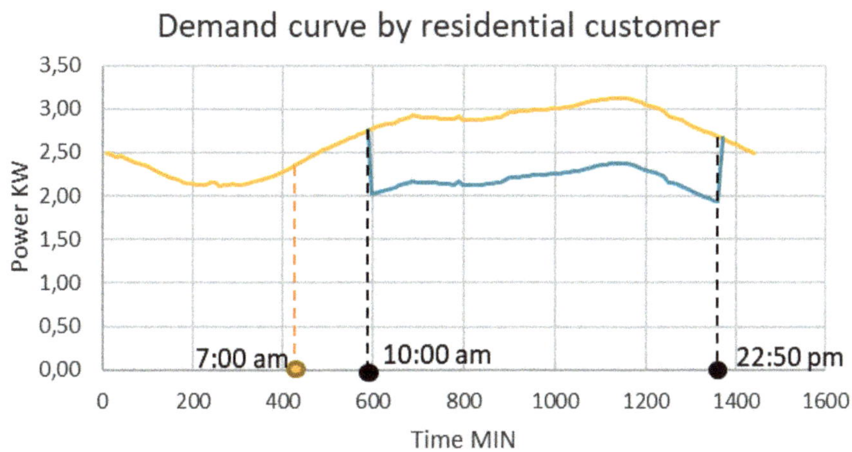

Fig. 8 Case 3: 750 W inverter with storage. Provides energy from 10h00 to 22h50

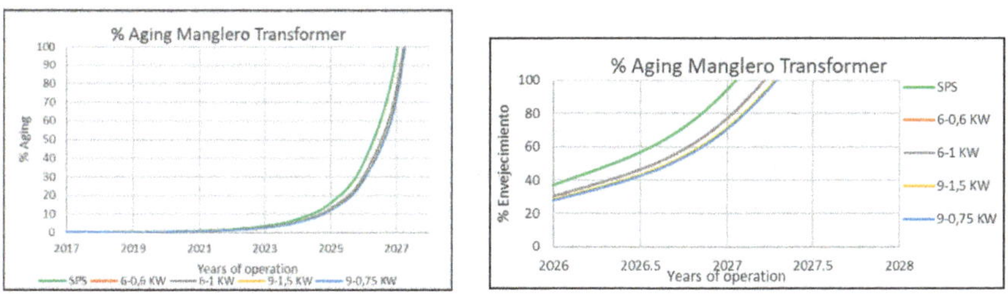

a b

Fig. 9 Modified aging curve Manglero transformer: (**a**) total curve, (**b**) zoom over end of lifetime

4.2 *Mocoli Option*

Several options were tested for the Mocoli substation, assuming that all subscribers supplying this substation would have the same solar generation system. Similar to Manglero, peak demand occurs at night. Therefore, a distributed storage system (batteries) [10] was implemented for those hours. As shown in Fig. 10, implementing the previously detailed system reduces transformer aging. Consequently, the service life will increase by approximately 5 months using 9 panels and one inverter of 0.75 KW per subscriber (Table 7 and Fig. 11).

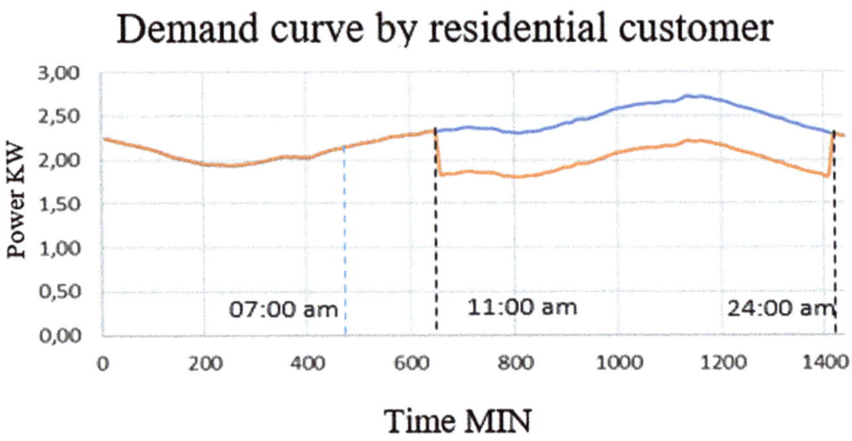

Fig. 10 Case 1: 500 W inverter with storage. Provides energy from 11h00 to 24h00

Table 7 Utilization factor reduction

Case	# Panels	Inverter	Utilization factor
Case 1	**6 panels**	**500 W**	**0.0930**
Case 2	6 panels	600 W	0.0950
Case 3	9 panels	750 W	0.0939
Case 4	9 panels	1000 W	0.0949
No panels			0.1088

5 Conclusions

The impact on the transformer is small as the exponential behavior of aging curve is not observable at the beginning of its life curve, resulting in a decrease of 3 and 4 °C for the Manglero and Mocoli transformers, respectively. This is reflected in a small lifespan of the equipment during the most part of its lifetime, as the exponential factor increases rapidly on a small timeframe.

The design and implementation of an equivalent 2.98 MW and 0.91 MW centralized solar generation system connected at Manglero and Mocoli substations, respectively, is not feasible, technically, or economically. The transformer's lifespan increases by 4 and 5 months, respectively, and the investment would exceed four million dollars compared to the approximately $300,000 that a transformer of the same capabilities' costs.

On the other hand, implementing a distributed storage system would be beneficial at the technical level for the distribution company, storing energy during valley hours to provide it at peak times and linearize the curve, without incurring high costs for economical battery systems, but as independent prosumers can generate a higher amount of energy, it is possible to increase the impact over each element of the grid and can have a better impact at long-term on lifetime of transformers.

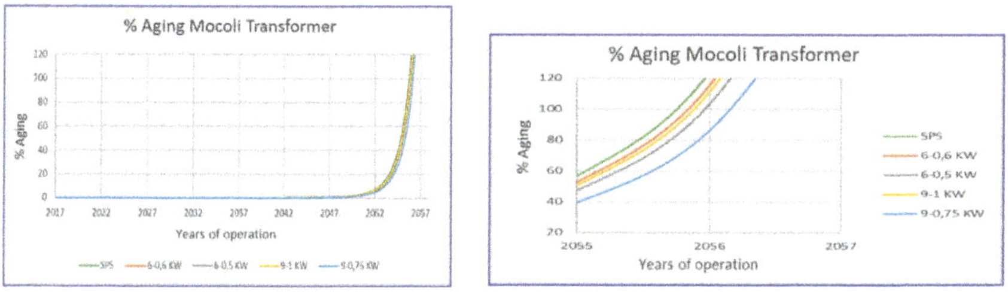

a b

Fig. 11 Modified aging curve Mocoli transformer: (**a**) total curve, (**b**) zoom over end of lifetime

Given the considerable costs of a residential solar generation system with storage, there should be implemented some load-shifting methods to incentive the demand behavior to shift from peaks to valley hours using storage systems.

References

1. IEEE, «IEEE Std C57.91-2011,» IEEE, New York, 2012.
2. IEEE STD. ASSOCIATION, IEEE Guide for Loading Mineral-oil-immersed Transformers and Step-Voltage Regulators, New York, USA: IEEE, 2012.
3. M. Koch, «Reliable Moisture determination,» Cigré, 2008.
4. L. Lundgaard et al., «Aging of oil-impregnated paper in transformers,» USA, 2002.
5. A. VILLA, «Estandarización e implementación de la norma ASTM D-4243 para la determinación del grado de Polimerización del papel eléctrico como asislante en transformadores de potencia por el método del promedio viscosimétrico.,» UNIVERSIDAD TECNOLÓGICA DE PEREIRA, Pereira.
6. CONSEJO NACIONAL DE ELECTRICIDAD, «Plan Nacional de Electrificacion del Ecuador,» Ecuador, 02/2002.
7. Gobierno Autonomo Descentralizado Municipal del Canton Samborondon, «Plan Cantonal de Desarrollo y Reordenamiento Territorial 2015-2019,» Samborondón, 2015.
8. Tamesol, «Tamesol,» 8/08/2017, https://www.tamesol.com/?lang=es, last accessed 12/082017.
9. Service my solar, «Service my solar,» 26/02/2015. Available: http://servicemy.solar/solar-inverter-repairs-and-service/omnik-inverter-repairs-and-service/. last accessed 17/07/2017.
10. Ulltracell, «Ultracell,» 14/12/2011. Available: http://www.ultracell.es/. last accessed 15/07/2017.

Assessment of Ocean Wave Energy Site Mapping in the Philippines

A Satellite Altimetry and SWAN Modeling Approach

Rexmelle Decapia Jr, Zachariah John Belmonte, Nemalyn Decapia, Louie Canta, and Princess Mary Angelique Beler

Abstract The Philippines emerged as a promising site for wave energy harvesting, considering significant wave height and wave period. A comprehensive assessment was conducted of potential ocean wave energy sites along the Philippine coastline specifically in Baler, Aurora by integrating data from Jason-2 and 3 satellite altimeters, SWAN model simulations, and in-situ measurements, utilizing Matlab and SWAN software. Rigorous validation methodologies were emphasized to ensure data accuracy through validation with in-situ measurements. The study revealed the influence of monsoon seasons on wave energy potential, offering insights into seasonal variations. Despite reduced energy potential during the westerly monsoon, significant wave energy during the easterly monsoon remained attractive. Maximum energy could be harnessed with a significant wave height of 2.5–3.0 m and a wave period of 7–8 s. These findings contribute to understanding site selection criteria, potential estimation, and methodologies in wave energy generation.

Keywords Ocean wave energy · Renewable energy · Site assessment · SWAN model

1 Introduction

Renewable energy plays a crucial role in the sustainable development of developing countries, offering a pathway to energy independence, environmental sustainability, and economic growth [1]. Ocean wave energy, a significant and untapped source of renewable energy, harnessed the power of ocean surface waves to generate electricity [2]. This form of energy was abundant, consistent, and had the potential to

R. Decapia Jr · Z. J. Belmonte (✉) · N. Decapia · L. Canta · P. M. A. Beler
Technological University of the Philippines – Taguig Campus, Taguig City, Philippines
e-mail: zachariahjohn_belmonte@tup.edu.ph

Z. Peng (ed.), *Technological Advancements and Future Directions in Green Energy*, Green Energy and Technology,
https://doi.org/10.1007/978-3-031-83203-1_3

significantly contribute to the global energy mix [1, 2]. Unlike solar or wind energy, which could be intermittent, ocean waves provide a more reliable and predictable energy source due to their continuous and cyclic nature [3]. In recent years, energy generated from ocean waves has been captured using various emerging technologies [2].

Developing countries like the Philippines often face significant challenges in meeting their energy demands [4]. In many regions, the energy infrastructure was underdeveloped, leading to limited access to reliable and affordable electricity. This energy gap hampered economic development and affected the quality of life. Traditional fossil fuel-based energy sources were not only depleting but also contributed to environmental degradation and climate change [5]. Therefore, there was a pressing need for sustainable and renewable energy solutions to bridge the energy challenges in the Philippines [6].

The Philippines, an archipelago with an extensive coastline, presented a unique opportunity for the development of ocean wave energy [3]. The country's energy mix plan aimed to diversify its energy sources, reduce dependence on imported fossil fuels, and enhance energy security. Incorporating renewable energy, particularly ocean wave energy, into the energy mix could provide a sustainable and locally sourced solution to the country's growing energy demands [7]. By leveraging its geographical advantages, the Philippines could become a leader in renewable energy adoption in the region. Recognizing this potential, the Energy Plan of the Philippines has begun to actively consider renewable energy resources [8].

However, one of the main constraints in energy planning in these regions was the poor data availability and reliability, as highlighted by Debnat et al. [9] and Riva et al. [10]. To address this issue, countries like the Philippines needed to develop a comprehensive database of various renewable energy sources. Providing studies for the assessment of ocean wave energy analysis would enable more accurate energy planning, policy-making, and investment in renewable energy infrastructure [8–10]. Sepula et al. [11] and Ribal et al. [12] utilized Jason-2 & 3 satellite altimeter data in ocean wave analysis. This data was complemented by simulating wave conditions nearshore using the Simulating WAves Nearshore (SWAN) model, which provided detailed insights into wave behavior and energy potential in coastal areas [11].

Additionally, in-situ measurements from various locations were used to validate and enhance the accuracy of the satellite and model data [13]. The use of the SWAN model represented a novel approach to assessing wave energy potential in the Philippines. The SWAN model was specifically designed to simulate wave propagation and transformation in coastal regions [11, 13], making it highly suitable for the complex and varied coastline of the Philippines. Its ability to provide high-resolution spatial data on wave energy distribution was a significant advancement over traditional methods, enabling more precise and localized energy resource assessments [10].

This study was the first to comprehensively examine the wave energy potential along the Philippine coastline. The general objective was to assess this potential by integrating data from Jason-2 and 3 satellite altimeters [9, 11], SWAN model simulations, and in-situ measurements [13]. To ensure the accuracy and reliability of the findings, satellite and model data were employed to validate with in-situ measurements [10–12].

2 Review of Related Literature

Figure 1 illustrates the conceptual framework of this study utilizing the Bathymetric data, Wind data, and Boundary Conditions utilizing the SWAN model by employing altimeter data (Jason 2 & 3) WTG measured Parameters to assess the Ocean Wave Energy in the Philippine Region specifically in Baler Coast.

2.1 Ocean Wave Energy

Assessing the feasibility and potential of harnessing ocean wave energy was crucial for its effective utilization, given its immense potential as a sustainable and abundant alternative to fossil fuels [3, 7, 11]. Research in the United States leveraged advanced modeling techniques and satellite data to assess wave energy potential [14]. Similarly, studies in the United Kingdom [15], Portugal [16], Australia [17], Japan [18], and Thailand [19] focused on methodologies for resource assessment and site selection. These efforts considered factors like wave climate variability and coastal dynamics. While challenges remained, ongoing advancements in methodologies, technologies, and interdisciplinary research efforts offered promising opportunities to improve assessment accuracy and reliability, making the path toward clean and renewable wave energy increasingly achievable [1, 5, 9].

2.2 Swan Model

Wave energy assessment was integral to understanding the viability and potential of wave energy as a renewable resource [3, 6]. Among the various methodologies employed, the Simulating WAves Nearshore (SWAN) model emerged as a valuable tool for assessing wave energy potential, providing detailed insights into wave behavior and characteristics. The SWAN model, along with remote sensing techniques such as satellite altimetry, enabled comprehensive assessments of wave energy resources by simulating wave propagation and transformation near coastlines [11, 13]. Its application

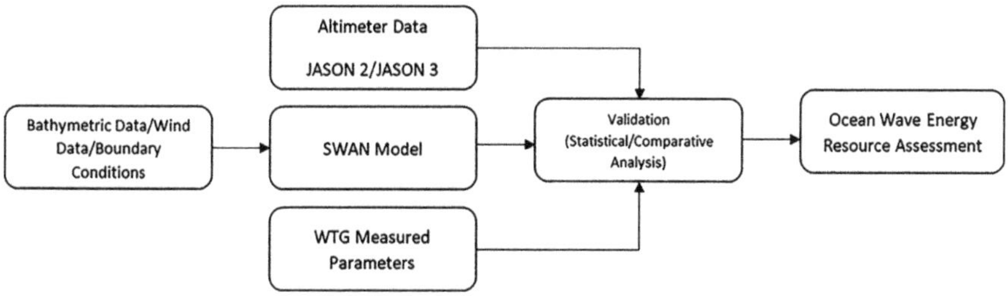

Fig. 1 Conceptual framework

significantly advanced our understanding of wave energy dynamics and was instrumental in identifying suitable sites for wave energy projects globally. Studies from various countries demonstrated the utilization of the SWAN model in wave energy assessment. In Europe [14], the United States [19], Australia [20], and the Asia-Pacific regions [21–23], the SWAN model was utilized to evaluate wave energy potential, identify optimal sites for wave energy installations, and optimize energy extraction strategies.

2.3 Data Collection

The University Ethics Committee reviewed and approved the process to ensure compliance with Republic Act 10173, also known as the Data Privacy Act of 2012 in the Philippines. Utilizing 10 years of a long-track Jason-2 and Jason-3 altimeter data collection has greatly advanced this study, determining the optimal timeframe [24, 30].

2.4 Sites of Assessment: Baler Aurora, Philippines Coasts

Baler, Aurora, Philippines, stands out as an ideal site for ocean wave energy due to its geographical advantages and favorable wave characteristics. Positioned along the Pacific Ocean, Baler is exposed to consistent and powerful wave activity, making it a prime candidate for wave energy exploration. Its extensive coastline and open ocean exposure facilitate the capture of significant wave energy, essential for generating renewable power. Baler's coastal configuration and seabed characteristics further enhance its suitability for wave energy projects. The Department of Energy (DOE) of the Philippines has identified Baler as a site with significant wave energy potential, experiencing considerable wave heights and periods crucial for wave energy generation. Significant wave height (SWH) refers to the average height of the highest third of waves, while wave period indicates the interval between successive wave crests. A study by Nioko et al. compared wave data from various sites in the Philippines and found that Baler consistently recorded higher wave heights and longer periods, making it more suitable for wave energy projects. The exact assessment site coordinates that are going to asses in this study are: Site 1 lat = 15.7882 lon = 121.575 and Site 2 lat = 15.8041 lon = 121.5583.

3 Methodology

Satellite altimeters, nadir-pointing instruments, were designed to precisely measure the time a radiated pulse took to travel to the surface and return [28]. They measured the time taken by a radar pulse to travel from the satellite antenna to the surface and back to the satellite receiver. Combined with precise satellite location data, altimetry measurements yielded sea-surface heights [29]. The Jason-2 satellite, launched on 20 June 2008, was the latest in a series of ocean altimeter missions designed to observe ocean circulation, sea level rise, and wave heights [30]. Information taken from these resources was validated through the collocation and concurrence method.

Additionally, the Simulating WAves Nearshore (SWAN), a third-generation wave modeling tool, was employed to interpret the data. SWAN used sophisticated mathematical equations and physics to predict individual wave heights and the expected wave field, which was the average of wave heights over a large area [10, 18]. The average commonly used by meteorologists was the significant wave height, defined as the average height of the highest one-third of all individual waves [15, 31]. This was also used to determine wind direction and speed. This wave model computed wave propagation, wave generation by wind, non-linear wave-wave interactions, and dissipation for a given bottom topography, wind field, water level, and current field in waters of deep, intermediate, and finite depth [14, 15, 25]. Moreover, input data for the generation model, such as bathymetry, was taken from the ETOPO1 Global Relief Model. Wave and wind data were taken from the Naval Global Environmental Model (NAVGEM NWW3), and boundary conditions data were generated from the Wavewatch III Model. Lastly, to validate and calibrate the model, in-situ measurements using wave-tide gauges deployed by the National Mapping and Resource Information Authority of the Philippines (NAMRIA) in various locations were used for comparison (Table 1).

Table 1 Tabulated methods of research with description

Methods	Description	Data source and software	References
Altimeter data	A 10-year along-track altimeter data of the significant wave height (SWH) was collected. The significant wave height seasonality and variability were analyzed, and the relationship between significant wave height, wind speed, and velocity was established. The data was sorted and processed using MATLAB for further analysis	Jason 2 and 3 Microsoft Excel Mathlab R2022b	[24] [28] [29]
Wave modelling	Developing the wave model using the SWAN software 20.1, estimating parameters such as average energy density, variability, seasonality, and directionality, creating a comprehensive wave resource map, and finally, identifying prime locations for wave energy harnessing	SWAN software 20.1	[2, 12] [16, 17]
In-situ measurements using wave-tide gauge	Selection of potential sites through the wave model, gathering comprehensive data from wave measurement projects, recording and processing data using MATLAB, collocating data points for consistency, and conducting a rigorous comparative analysis among altimeter data, the SWAN model, and in-situ measurements. This approach ensures a thorough examination and validation of wave energy along the coastline	NAMRIA Microsoft Excel Mathlab R2022b	[10, 11] [12, 13] [22]

4 Results and Discussion

Figure 2 shows the comparison between the significant wave height from the raw WTG data and the result from SWAN simulation. It is observable that the raw WTG data is affected by the tidal change in the ocean. To minimize the effect of the tidal movement, the raw WTG data was denoised using the Fast Fourier transform [25, 30].

The correlation value was increased from 0.74 to 0.76 which indicates a strong positive correlation [28].

Additionally, Fig. 3 defined the comparison between the wave period from the raw WTG data and the result from the SWAN simulation. It is observable that the raw WTG data is affected by the tidal change in the ocean. To minimize the effect of the tidal movement, the raw WTG data was denoised using the Fast Fourier transform [25, 30]. The correlation value was increased from 0.50 to 0.56 which indicates positive moderate correlation [28].

Moreover, Fig. 4 illustrates the monthly average significant wave height in meters in the area covered. The data shows that its value decreases at the start of the westerly monsoon (habagat) during March to September and starts to increase significantly during the easterly monsoon (amihan) during the months of October to February. This indicates a favorable opportunity for wave energy harvesting during the amihan.

Subsequently, Fig. 5 shows a similar trend, indicating high wave periods during the easterly monsoon which resulted in a higher wave energy potential during the season as shown in Fig. 5. A significant amount of energy can be harvested during October to January.

To further assess the ocean wave energy potential, two additional sites were evaluated, with considerations near the shore taken into account as a criterion. The data (shown in Fig. 6) revealed a change in significant wave height during the eastern monsoon, along with a slight change in wave period (as depicted in Fig. 7). This is

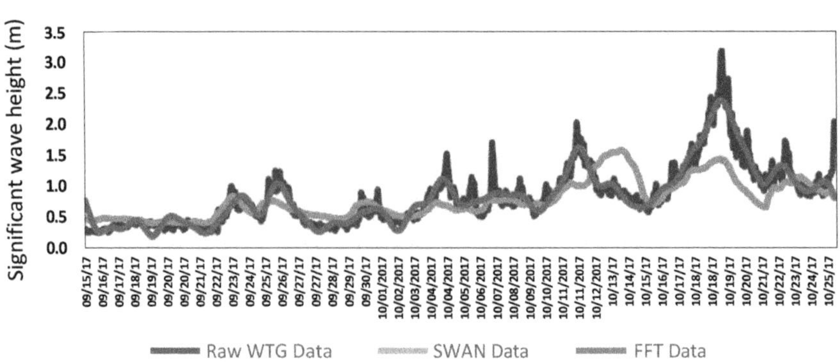

Fig. 2 Comparison of the significant wave heights from the raw WTG data, swan simulation, and the denoised WTG data using FFT from September 12, 2017 (16:00) to October 26, 2017 (9:00)

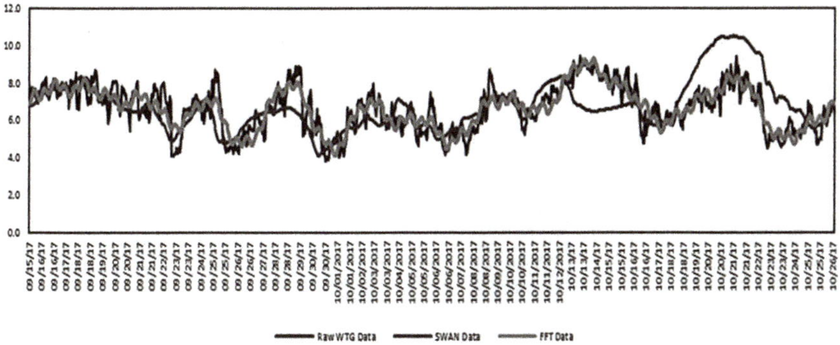

Fig. 3 Comparison of the significant wave periods from the raw WTG data, swan simulation, and the denoised WTG data using FFT from September 12, 2017 (16:00) to October 26, 2017 (9:00)

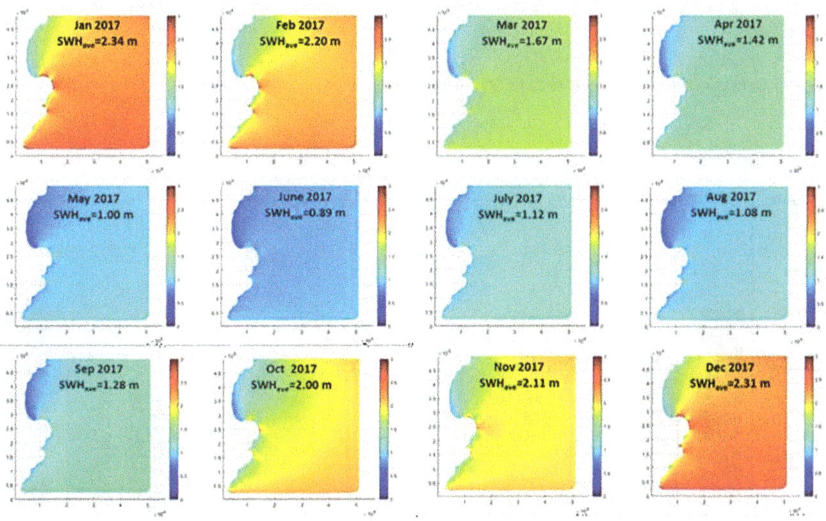

Fig. 4 Average monthly significant wave height (m) in the area considered

primarily attributed to the fact that the nearshore conditions have minimal influence on the wave cycle, with wind becoming the predominant factor affecting wave height [25, 26, 28]. Moreover, the analysis indicated that the trend for wave energy at the selected site correlates to the findings observed across the entire area, as illustrated in Figs. 7, 8, and 9.

The estimated average annual wave energy potential in site 1 is around 1, 975 kW/m in which a large portion of this energy generated by a wave period of 7–8 s (approximately 51.5%) while the significant wave height of 2.5 – 3.5 m contributed to around 38.4% of the total estimated energy. This data showed a favorable result since a higher probability will also ensure higher reliability (Fig. 10).

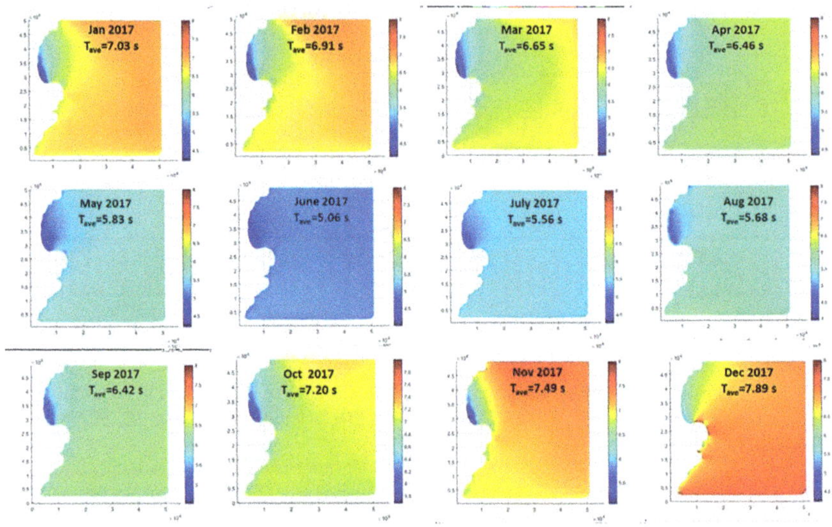

Fig. 5 Average monthly wave period (s) in the area considered

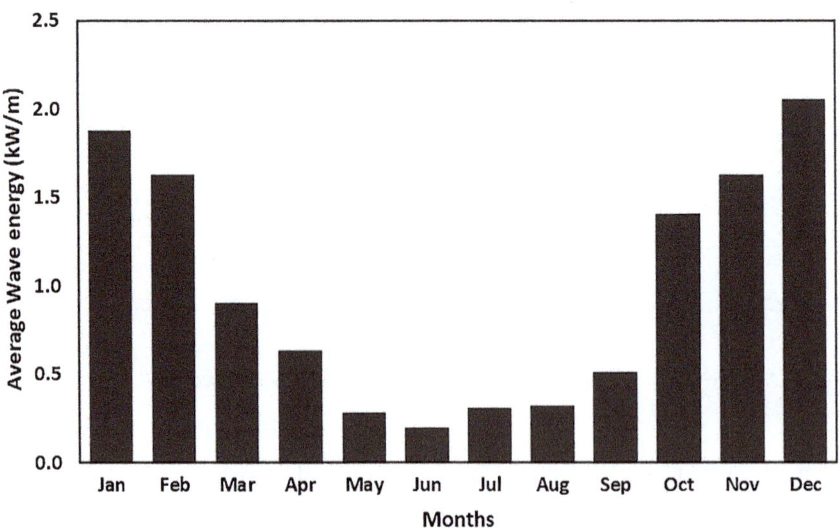

Fig. 6 Average monthly wave energy in kW/m

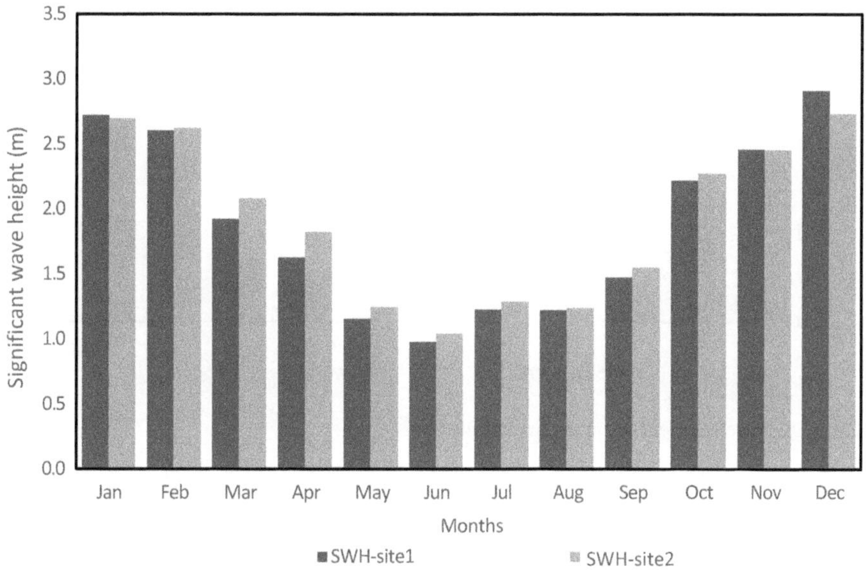

Fig. 7 Average significant wave height in the two selected sites

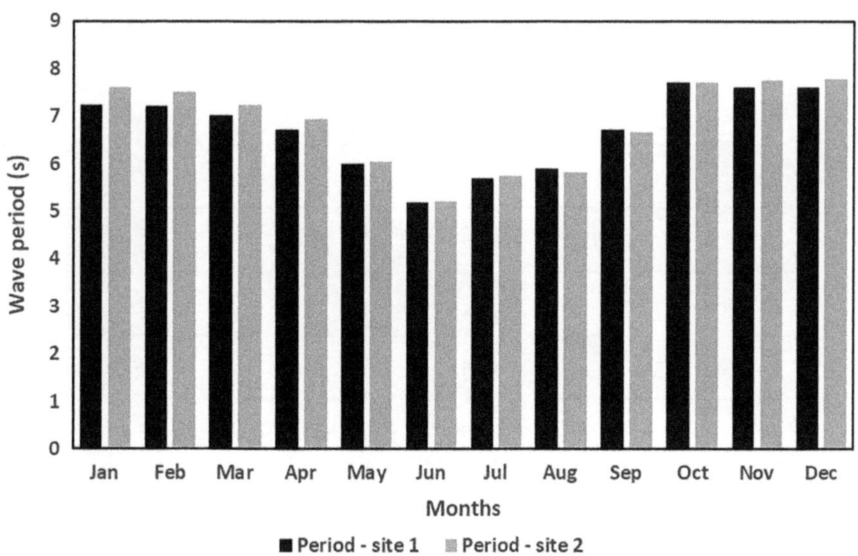

Fig. 8 Average wave period in the two selected sites

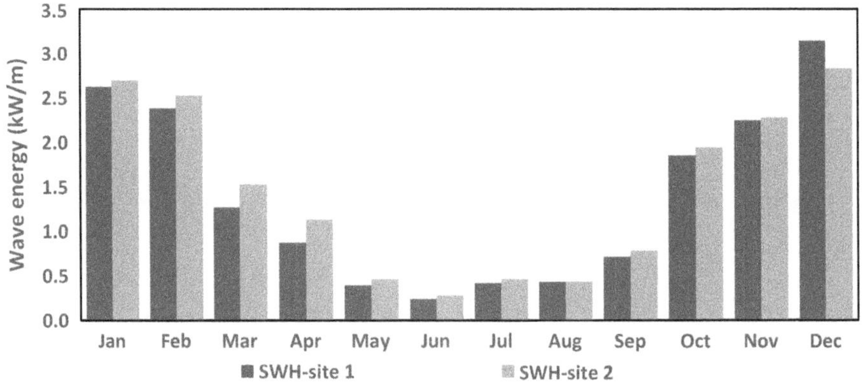

Fig. 9 Average wave energy in the two selected sites

SWH/ Period	4-5	5-6	6-7	7-8	8-9	9-10	Total
4-4.5				4.5%	2.1%		6.6%
4.5-5				0.4%	1.3%		1.7%
3-3.5			1.5%	[illegible]	1.2%	2.3%	18.7%
3.5-4			0.5%	4.7%	2.3%		7.5%
2-2.5		0.3%	4.9%	10.9%	3.0%		19.2%
2.5-3			6.5%	10.2%	3.3%	0.5%	20.8%
1-1.5	0.3%	3.3%	4.0%	1.1%	0.2%		8.9%
1.5-2		0.8%	4.4%	7.0%	2.2%	0.1%	14.4%
0.5-1	0.7%	1.1%	0.3%				2.1%
Total	1.0%	5.5%	21.9%	51.5%	15.7%	2.9%	100.0%

Fig. 10 Energy matrix for site 1

For site 2, the estimated average annual wave energy potential is around 6641 kW/m in which a large portion of this energy is generated by a wave period of 7–8 s (approximately 59.0%) while the significant wave height of 2.5–3.5 m contributed to around 44.67% of the total estimated energy. This data showed a favorable result since a higher probability will also ensure higher reliability (Fig. 11).

4.1 Practical Implication and Managerial Contributions

This study offers practical insights for wave energy project development in the Philippines, Baler, Aurora as a potential site for Ocean Wave Energy. By analyzing seasonal variations and monsoon impacts, it guides operational strategies for efficient energy generation. Stakeholders can use these findings to make informed decisions on infrastructure deployment and management. The study's methodology provides a framework for future assessments in coastal regions and other countries.

SWH/ Period	4-5	5-6	6-7	7-8	8-9	9-10	Total
7.5-8				3.6%			3.6%
5-5.5							0.0%
3-3.5				8.0%	4.0%	2.2%	14.9%
3.5-4					3.8%		3.8%
2-2.5			3.6%	17.5%	1.8%	0.2%	23.0%
2.5-3			0.4%	19.1%	2.9%	0.4%	23.2%
1-1.5		5.8%	6.3%	1.6%			13.8%
1.5-2		0.3%	2.6%	9.1%	2.1%		14.2%
0.5-1	0.7%	2.4%	0.3%				3.4%
Total	0.7%	8.6%	13.3%	59.0%	14.5%	2.7%	100.0%

Fig. 11 Energy matrix for site 2

4.2 Theoretical Implications

This study pioneers a comprehensive assessment of wave energy potential along the Philippine coastline, integrating data from Jason-2 and 3 satellite altimeters, SWAN model simulations, and in-situ measurements [27, 28, 30]. By rigorously validating satellite and model data with in-situ measurements, it identifies Baler, Philippines, as a promising site for wave energy harvesting, considering factors such as significant wave height and wave period. Insights into seasonal variations and the influence of monsoon seasons on wave energy potential are also provided. These findings contribute to site selection criteria, potential estimation methodologies, and temporal dynamics in wave energy generation, advancing research and development in wave energy projects.

5 Conclusions

By integrating data from Jason-2 and 3 satellite altimeters, SWAN model simulations, and in-situ measurements, this study assessed the potential ocean wave energy sites in the Philippines. Emphasizing the importance of rigorous validation methodologies, the research ensured the accuracy of satellite and model data through validation with in-situ measurements. Moreover, it identified Baler, Philippines, as a promising site for wave energy harvesting, considering factors such as significant wave height and wave period. Additionally, the study explored the influence of monsoon seasons on wave energy potential, providing insights into seasonal variations. These contributions advanced understanding in site selection criteria, potential estimation methodologies, and temporal dynamics in wave energy generation, offering valuable insights for future research and development in wave energy projects. Despite the decrease in potential energy during the westerly monsoon, the significant wave energy during the easterly monsoon remains attractive. It has been found that the maximum energy can be harnessed with a significant wave height of 2.5–3.0 m and a wave period of 7–8 s.FundingThis research was funded by the Technological University of the Philippines and reviewed and approved by the University Ethics Committee.

References

1. Dincer, Ibrahim. "Renewable energy and sustainable development: a crucial review." Renewable and sustainable energy reviews 4, no. 2 (2000): 157–175.
2. Belmonte, Zachariah John A., Yogi Tri Prasetyo, Omar Paolo Benito, Jui-Hao Liao, Krisna Chandra Susanto, Michael Nayat Young, Satria Fadil Persada, and Reny Nadlifatin. "The acceptance of nuclear energy as an alternative source of energy among Generation Z in the

Philippines: An extended theory of planned behavior approach." Nuclear Engineering and Technology 55, no. 8 (2023): 3054–3070.
3. Neill, Simon P., and M. Reza Hashemi. Fundamentals of ocean renewable energy: generating electricity from the sea. Academic Press, 2018.
4. Belmonte, Zachariah John, Charles Bandola, Rexmelle Decapia, and Eugene Chris Gonzaga. "Do technological developments reduce unemployment in the Philippines?." In 2021 62nd International Scientific Conference on Information Technology and Management Science of Riga Technical University (ITMS), pp. 1–5. IEEE, 2021.
5. Capellán-Pérez, Iñigo, Margarita Mediavilla, Carlos de Castro, Óscar Carpintero, and Luis Javier Miguel. "Fossil fuel depletion and socio-economic scenarios: An integrated approach." Energy 77 (2014): 641–666.
6. Salac, Arizeo C., Jairus Dameanne C. Somera, Michael T. Castro, Maricor F. Divinagracia-Luzadas, Louis Angelo M. Danao, and Joey D. Ocon. "Off-Grid Electrification Using Renewable Energy in the Philippines: A Comprehensive Review." Smart Cities 7, no. 3 (2024): 1007–1043.
7. Cherp, Aleh, Adeola Adenikinju, Andreas Goldthau, Francisco Hernandez, Larry Hughes, Jessica Jewell, Marina Olshanskaya, Jaap Jansen, Ricardo Soares, and Sergey Vakulenko. "Energy and security." In Global energy assessment: Toward a sustainable future, pp. 325–383. Cambridge University Press, 2012.
8. Mamat, Rizalman, Mohd Shahrir Mohd Sani, Fitri Khoerunnisa, and Asep Kadarohman. "Target and demand for renewable energy across 10 ASEAN countries by 2040." The Electricity Journal 32, no. 10 (2019): 106670.
9. Debnath, Kumar Biswajit, and Monjur Mourshed. "Challenges and gaps for energy planning models in the developing-world context." Nature Energy 3, no. 3 (2018): 172–184.
10. Riva, Fabio, Annalisa Tognollo, Francesco Gardumi, and Emanuela Colombo. "Longterm energy planning and demand forecast in remote areas of developing countries: Classification of case studies and insights from a modelling perspective." Energy strategy reviews 20 (2018): 71–89.
11. Sepulveda, Hector Hito, Pierre Queffeulou, and Fabrice Ardhuin. "Assessment of SARAL/AltiKa wave height measurements relative to buoy, Jason-2, and Cryosat-2 data." Marine Geodesy 38, no. sup1 (2015): 449–465.
12. Ribal, Agustinus, and Ian R. Young. "33 years of globally calibrated wave height and wind speed data based on altimeter observations." Scientific data 6, no. 1 (2019): 77.
13. Balsamo, Gianpaolo, Anna Agusti-Parareda, Clement Albergel, Gabriele Arduini, Anton Beljaars, Jean Bidlot, Eleanor Blyth et al. "Satellite and in situ observations for advancing global Earth surface modelling: A review." Remote Sensing 10, no. 12 (2018): 2038.
14. Vos, Kilian, Mitchell D. Harley, Kristen D. Splinter, Andrew Walker, and Ian L. Turner. "Beach slopes from satellite-derived shorelines." Geophysical Research Letters 47, no. 14 (2020): e2020GL088365.
15. Jacobson, Paul T., George Hagerman, and George Scott. Mapping and assessment of the United States ocean wave energy resource. No. DOE/GO/18173-1. Electric Power Research Institute, 2011.
16. Mota, P., and J. P. Pinto. "Wave energy potential along the western Portuguese coast." Renewable Energy 71 (2014): 8–17.
17. Hemer, Mark A., Stefan Zieger, Tom Durrant, Julian O'Grady, Ron K. Hoeke, Kathleen L. McInnes, and Uwe Rosebrock. "A revised assessment of Australia's national wave energy resource." Renewable Energy 114 (2017): 85–107.
18. Webb, Anthony, T. Waseda, and K. Kiyomatsu. "A high-resolution, long-term wave resource assessment of Japan with wave–current effects." Renewable energy 161 (2020): 1341–1358.
19. Kompor, Wongnarin, Chaiwat Ekkawatpanit, and Duangrudee Kositgittiwong. "Assessment of ocean wave energy resource potential in Thailand." Ocean & coastal management 160 (2018): 64–74.

20. Morim, Joao, Nick Cartwright, Amir Etemad-Shahidi, Darrell Strauss, and Mark Hemer. “A review of wave energy estimates for nearshore shelf waters off Australia.” International Journal of Marine Energy 7 (2014): 57–70.
21. Kamranzad, Bahareh, and Pengzhi Lin. “Sustainability of wave energy resources in the South China Sea based on five decades of changing climate.” Energy 210 (2020): 118604.
22. Sasmal, Kaushik, Takuji Waseda, Adrean Webb, Shogo Miyajima, and Kunio Nakano. “Assessment of wave energy resources and their associated uncertainties for two coastal areas in Japan.” Journal of Marine Science and Technology 26 (2021): 917–930.
23. Eum, Ho-Sik, Weon-Mu Jeong, Yeon S. Chang, Sang-Ho Oh, and Jong-Jip Park. “Wave energy in korean seas from 12-year wave hindcasting.” Journal of Marine Science and Engineering 8, no. 3 (2020): 161.
24. Ma, Shanmu, Jingjuan Liao, Ruofan Jing, and Jiaming Chen. “A dataset of lake level changes in China between 2002 and 2023 using multi-altimeter data.” Big Earth Data (2024): 1–23.
25. Nioko, L. I. J., K. M. C. Tablang, A. M. Tamondong, M. R. C. O. Ang, and D. M. Bautista. “TYPHOON KARDING (NORU) STORM SURGE ANALYSIS USING THE COAWST MODELING SYSTEM." The International Archives of the Photogrammetry, Remote Sensing and Spatial Information Sciences 48 (2024): 403-410.
26. Bertram, D. V., A. H. Tarighaleslami, M. R. W. Walmsley, M. J. Atkins, and G. D. E. Glasgow. “A systematic approach for selecting suitable wave energy converters for potential wave energy farm sites.” Renewable and Sustainable Energy Reviews 132 (2020): 110011.
27. Farkas, Andrea, Nastia Degiuli, and Ivana Martić. “Assessment of offshore wave energy potential in the Croatian part of the Adriatic Sea and comparison with wind energy potential.” Energies 12, no. 12 (2019): 2357.
28. Fu, Lee-Lueng, and Anny Cazenave, eds. Satellite altimetry and earth sciences: a handbook of techniques and applications. Elsevier, 2000.
29. Martin, Scott A., Georgy E. Manucharyan, and Patrice Klein. “Synthesizing sea surface temperature and satellite altimetry observations using deep learning improves the accuracy and resolution of gridded sea surface height anomalies.” Journal of Advances in Modeling Earth Systems 15, no. 5 (2023): e2022MS003589.
30. Lambin, Juliette, Rosemary Morrow, Lee-Lueng Fu, Josh K. Willis, Hans Bonekamp, John Lillibridge, Jacqueline Perbos et al. “The OSTM/jason-2 mission.” Marine Geodesy 33, no. S1 (2010): 4–25.
31. Feng, Xiangbo, M. N. Tsimplis, G. D. Quartly, and M. J. Yelland. “Wave height analysis from 10 years of observations in the Norwegian Sea.” Continental Shelf Research 72 (2014): 47–56.

Part II
Environmental Impact Assessment of Hydrogen Energy Production and Environmental Pollution Management

Comparative Life Cycle Analysis of Electrolysis Hydrogen Production from Diverse Green Energy Sources

Wagd Ajeeb, Rui Costa Neto, and Patrícia Baptista

Abstract Green hydrogen (GH_2) has been considered a promising alternative to fossil fuels in chemical and energy applications. In this study, a comparative analysis is performed for the life cycle of GH_2 production, through pressurized Alkaline electrolysis (ALE), sourced with different renewable energy options in Portugal, especially hydropower, PV solar, and wind energy. GWP impact is assessed using SimaPro software concerning the construction materials and operation requirements for the hydrogen production project with 15 MW capacity and 20 years lifetime. Four electricity supply technology scenarios for the electrolyzer system are investigated, namely, hydropower, wind turbine, PV solar, and the electricity grid mix in Portugal. LCA results demonstrated that hydrogen from the hydropower scenario has the lowest GWP impact of 0.26 kg CO_2eq /kgH_2, followed by 1.76 for the wind turbine scenario, followed by 2.94 for the PV solar scenario, and 7.4 kg CO_2eq /kgH_2 for the grid mix scenario. The GWP values obtained are lower than the common conventional steam methane reforming method (11.89 kg CO_2eq /kgH_2).

Keywords Green H_2 · Electrolyzer · LCA · GWP · Green energy

W. Ajeeb (✉) · R. C. Neto
IN+ Center for Innovation, Technology and Policy Research, IST-ID -Associação do Instituto Superior Técnico para a Investigação e Desenvolvimento, University of Lisbon, Lisbon, Portugal
e-mail: wagdajeeb@tecnico.ulisboa.pt

P. Baptista
IN+ Center for Innovation, Technology and Policy Research, Instituto Superior Técnico, University of Lisbon, Lisbon, Portugal

Z. Peng (ed.), *Technological Advancements and Future Directions in Green Energy*, Green Energy and Technology,
https://doi.org/10.1007/978-3-031-83203-1_4

1 Introduction

The pursuit of reducing greenhouse emissions and transitioning away from fossil fuels has become a focal point for the global industrial community [1]. Renewable energy resources (RES) like hydro, wind, and solar power are increasingly prevalent, continuously advancing in efficiency, storage capabilities, and cost reduction [1, 2]. These RES electricity sources can be harnessed to produce green hydrogen through the process of electrolysis [3], facilitating the creation of chemical products and biofuels essential for various applications [4, 5]. Hydrogen, generated through electrolysis, holds promise as a versatile energy carrier for energy storage, power conversion, and transportation applications [6, 7], offering a carbon-neutral energy option [8]. As RES technologies mature and the levelized cost of energy (LCOE) decreases, they become increasingly competitive with fossil fuels [1, 9]. The convergence of advancements in hydrogen production methods and RES technologies presents an opportunity for their integration, fostering the sustainable production of green hydrogen and a cleaner environment devoid of fossil fuel dependence [10, 11]. In this context, it is an opportune time to leverage RES for green hydrogen production, capitalizing on its environmental benefits [12]. This trend is evident in Portugal, where the European Union's roadmap for Carbon Neutrality (RNC 2050) and the National Energy and Climate Plan (NECP) outline pathways for decarbonization, positioning hydrogen as a versatile form of green energy produced through various methods [13]. Despite the commercial feasibility of hydrogen production from fossil fuels, notably through steam methane reforming (SMR)) [14], these methods pose significant environmental challenges [15]. Notably, the greenhouse gas (GHG) emissions associated with SMR-based hydrogen production are significant, with estimates ranging from 11.7 to 11.89 kg CO_2-eq/kg H_2, primarily attributable to the operational requirements of the system [16, 17]. Koj et al. [18] observed that the environmental impacts of a 6 MW ALEs system were notably influenced by the location's electricity mix supply conditions, with Austria demonstrating superior environmental performance compared to Spain and Germany. Despite the environmentally friendly operation of hydrogen electrolysis methods powered by RES, the construction and production of components for electrolysis systems and renewable plants continue to pose environmental challenges [19].

Therefore, careful assumptions are crucial for LCA studies on hydrogen production, especially when key parameters like site conditions, lifecycle duration, materials, and energy consumption are lacking. A precise analysis of hydrogen electrolysis production technologies with different RES types (e.g., hydro, wind, PV solar) can guide the optimization of operational procedures and materials, improving the environmental performance (e.g., GWP impact), overall efficiency, and resource management while enhancing economic and social benefits [20]. Thus, to the authors' knowledge, the comparative LCA of the multi electricity supply scenarios to ALE technology for hydrogen production in Portugal addressed in this study is novel and there is no similar study in the literature. In addition, the comprehensive LCA of GH_2 production is performed for specific system boundaries selected in the present study, e.g., the site in Portugal, a capacity of 15 MW, electricity supply, manufacturing and importing the ALE from Northern Europe.

2 Goal and Scope

The modelling for this study is conducted by SimaPro software, gathering data from various confidential sources found in literature, project sites, company producers, expert insights, and the SimaPro inventory. Adhering to ISO 14044 requirements [21] and relevant guidelines for hydrogen production systems [22], the conducted methodology encompasses four stages: goal and scope definition, inventory analysis, impact assessment, and interpretation. The calculation process depends on the International Reference Life Cycle Data (ILCD) following the European Commission recommendations [23]. The ILCD 2011 Midpoint provided by SimaPro is considered a proper evaluation method that applies characterization factors for impact evaluation including GWP, using impact assessment models e.g., EC-JRC.

Firstly, this study aims to fill the research gab and identify the most suitable and environmentally friendly RES technology to feed the electrolysis hydrogen production in Portugal, specifically comparing hydro, PV solar, and wind supply technologies. The hydrogen production through the electrolysis phase is examined, and four electricity supply scenarios are compared, namely, the hydropower, PV solar, wind turbines, and the national electricity grid mix option. Also, Portugal was chosen as the case study location due to its significant progress in implementing RES in electricity generation. Presently, Portugal's electricity production comprises a mix of energy resources, including fossil fuels and RES, with a notable shift towards increased RES utilization particularly wind and solar energy resources. These RES offer diverse environmental advantages to the country. Consequently, the study analyzes the GWP impact of the hydrogen production associated with each RES technology scenario within identical system boundaries for the electrolyzer system and the same project capacity of 15 MW. The assessment focuses on the production of 1 kg of hydrogen, with specific conditions including a pressure of 33 bar, a temperature of 40 °C, and a purity of 99.8%.

3 Life Cycle Inventory

The operation boundaries of the water electrolysis system are important, encompassing both the inputs (materials and processes) and process stages as clarified in Fig. 1. The construction stage involves the two main units: the energy source system (hydropower plants, PV solar panels, wind turbines, or the grid electricity mix) and water electrolysis system. Each unit encompasses raw materials acquisition, manufacturing, transportation, operation and maintenance, dismantling, and waste

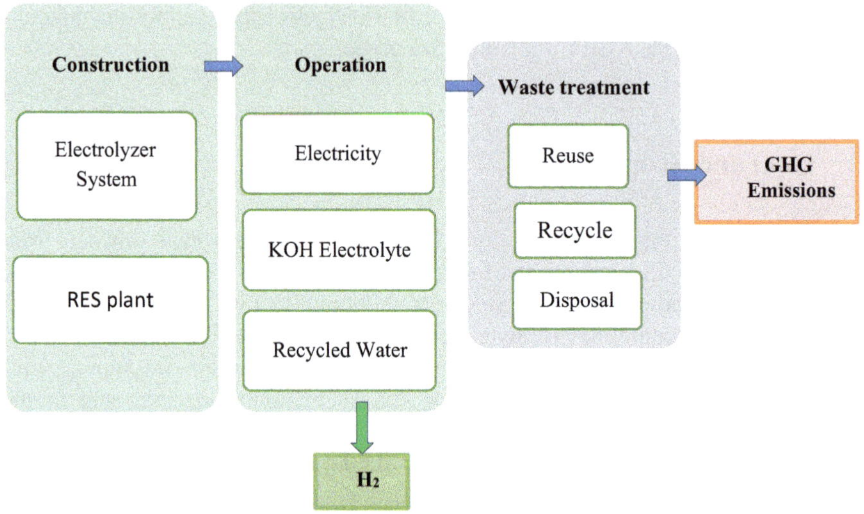

Fig. 1 A scheme for the stages and components of the life cycle for ALE hydrogen production

treatment. Through inventory analysis using SimaPro software (version 9.3.0.3), the defined inputs and outputs from the electrolysis process were assessed and categorized based on GWP impact. Data utilized in this LCA study were obtained from producer companies, experts, and relevant literature, and integrated into the SimaPro database (e.g., power supply, materials, and processes). The operation inputs data for both the RES plant and electrolyzer system including water, potassium hydroxide (KOH), cleaning materials, maintenance requirements, weather and industrial conditions (e.g., solar irradiation, and wind speed), are mainly sourced from the Portuguese database. In cases where Portuguese data is unavailable, European databases are considered for evaluation.

3.1 Electrolyzer System

The used electrolyzer in the current study is a pressurized ALE type which typically consists of two Nickel electrodes separated by a membrane layer and an Alkaline electrolyte liquid [24]. At the cathode side, hydroxide ions (OH−) are generated and migrate through the membrane to the anode side, where hydrogen is produced [25]. Alkaline electrolysis operates at relatively low temperatures, typically ranging from 30 to 80 °C, and utilizes readily available and cost-effective materials for its electrolyte solution, membrane, and electrodes [26, 27]. A recent model of ALE, designed for high-pressure operation (pressurized), has demonstrated promising performance, particularly in applications requiring compressed hydrogen [28, 29]. For a hydrogen project capacity of 15 MW, a commercial model was selected, comprising 10 stacks, each containing 139 cells [30]. This model is specified to be produced in Northern Europe. Relevant information, particularly regarding the mass of materials used in ALE production and operational requirements, was obtained from producer companies and available data for lower-capacity projects, such as 6 MW ALEs, were upscaled and utilized [18]. Furthermore, the characteristics of the pressurized ALE have been reported by several research studies (an efficiency of 80.05% HHV, for the ALE was defined) and commercial documentation [18, 31]. According to the information collected for ALE, common materials such as nickel, steel, aluminium, copper, and rubber are used for the electrolyzer components in the cell frame, gasket, and cathodes, as well as the Zirfon Membranes. Also, alongside the stack components, the Balance of Plant (BoP) is taken into account in the current LCA. This includes tanks, heat exchangers such as boiling heat exchanger [32, 33] or compact/micro heat exchangers [34, 35] and enhanced thermal fluid [36], pumps, power electronics/inverters, potassium hydroxide filters, and gas separators.

In addition, operational considerations encompass the provision of deionized water and potassium hydroxide solution (25% KOH v/v). Deionized water is sourced from a nearby wastewater treatment plant as recycled water. A lifetime of 10 years is assumed for both the KOH and the ALE stacks, necessitating replacement upon reaching the end of their lifetime. The total lifetime of the hydrogen production system under study is set at 20 years, with an operational period of 8300 hours per year. It is estimated that the production of 1 kg of hydrogen consumes approximately 10 L of deionized water in the entire system, including the Balance of Plant (BoP). Additionally, steam and nitrogen are utilized for cleaning the ALE system during its operational phase.

3.2 *Electricity Supply and Operation*

This study involves powering the electrolysis system with four different scenarios, namely, the hydropower, wind turbines, solar (PV), and electricity grid mix in Portugal, as depicted in Fig. 2. The scenarios consider annual yield ranges and the

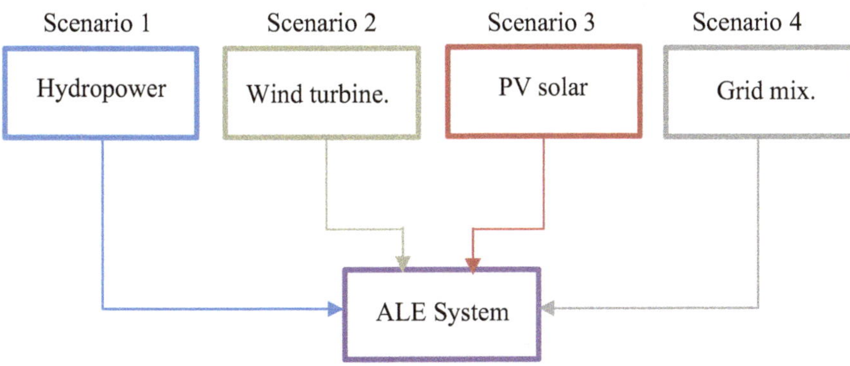

Fig. 2 A schematic for the energy scenarios to supply the ALE hydrogen production system

lifetime of the RES plants, with a total capacity of 15 MW from each power plant to operate the electrolysis process in the Alkaline stacks. Electricity production data from each RES type (hydropower, wind, and solar) are defined from the Ecoinvent version 3 database in SimaPro, reflecting the conditions and performance of each RES in Portugal. Site-specific parameters such as water flow, wind speed, and solar radiation, along with technology factors such as the lifetime of each RES technology are considered due to their significant influence on the environmental impacts of the overall system.

PV solar PV solar panels are selected in SimaPro database for low voltage electricity generation, with a 570 kWp open ground plant utilizing multi-Si PV panels. The plant incorporates an inverter to convert low-voltage DC into AC power and utilizes water for cleaning purposes.

Wind turbine The onshore wind turbines chosen for this study have a capacity of 4.5 MW each, particularly the Enercon E-112 type, which includes infrastructure elements such as rotor, tower, blades, and base foundation, as well as electricity grid components like voltage converters, connections, and cables [37].

Hydropower The hydropower plant was chosen in SimaPro database as a run-of-river power plant unit. Low-pressure power plants including river power stations and canal power plants are covered in the database. The determined specific data was then related to the entire power plant in Portugal. Lifetime is assumed to be 80 years for the structural part (including cement, reinforcing steel, diesel, electricity, transportation, and explosives) and 40 years for the rest of the materials. The turbine technology used has an overall efficiency of 88%. The efficiency of generators depends on the output, the rotation speed of the generator and the cooling system. For today's power plants, an efficiency of approximately 96% is assumed.

Electricity grid mix The electricity grid mix in Portugal is considered for the data collected in 2023 and modelled in SimaPro software, where it involves 92% of electricity generated in Portugal from different resources (27% wind, 10% solar, 30% hydropower, 8% bioenergy, 21% natural gas, and 4% other fossil fuels) [38] and 8% of electricity is imported from Spain [39].

However, in SimaPro, the inventory of hydropower, wind turbines, and PV panels are modelled for Portugal, taking into account the production of high voltage 1 kWh of electricity distributed to the grid, as well as operation and maintenance activities including the periodic replacement of materials.

3.3 Waste Treatment

Incorporating the recycling of electrolysis components, along with enhancing reuse rates and minimizing landfill disposal, should be accounted for any Life Cycle Assessment (LCA) investigation. In this study, the waste treatment (disposal equipment phase) methodology of hydrogen production by the ALE system follows the strategy presented by Lotric et al. [40]. The components of the system were considered to be manually dismantled. The rates of recycling for each material (steel, nickel, plastics, electronics) were estimated according to the recycling industry and available literature, for some materials such as 60% reuse of steel and 96% recycling of aluminium. Also, most other materials in the system were considered to have a 99% recycling rate and a 1% rate for landfills. The option of landfill disposal was selected solely in cases where material recovery or recycling was deemed unfeasible.

4 Results and Discussion

4.1 GWP Impact of Hydrogen Production

GWP impact has been considered a significant impact in LCA studies representing the GWG emissions. For the current study, the validation of the modelled RES types (the four scenarios) was validated primarily according to the GWP impact reported in related literature. Generally, GHG emissions from wind turbines typically vary from 8 to 62 g CO_2-eq per kWh, while PV systems can reach up to 132 g CO_2-eq per kWh [19]. A study conducted by Schleisner [41] on offshore and onshore wind turbines revealed GHG emissions data of 16.5 and 9.7 g CO_2-eq/kWh, respectively. Recent advancements in PV panel production technology have led to decreased GHG emissions of between 12 and 17.6 g CO_2 eq per kWh in the EU [42]. The hydropower, wind turbines and PV panels modelled in the current study exhibit GWP values of 1.16 g CO_2-eq/kWh for hydropower, 16.7 g CO_2-eq/kWh for PV, and 9.8 g CO_2-eq/kWh for wind turbines, aligning with the observed improvements reported in the literature. Figure 3 presents the results of GWP impact from hydrogen production for the four electricity supply scenarios.

The total GWP of the total LCA (considering all stages) for hydrogen production through ALE technology is demonstrated to be in the lowest value (0.26 kg CO_2-eq /kgH_2) when using the hydropower scenario, lower by −97% compared to the Grid mix scenario. Also, a GWP impact of 1.76 is found for the wind turbine scenario which is lower by −76% compared to the Grid mix scenario. Moreover, the GWP impact of 2.94 kg CO_2-eq /kgH_2 for the PV solar scenario is lower by −76% compared to the Grid mix scenario. Also, the Grid mix scenario exhibited a good GWP impact of 7.4 kg CO_2-eq /kgH_2 compared to the conventional SMR method for hydrogen production, due to the high percentages of multi-RES contributing to Portuguese grid electricity. The GWP values obtained are lower than the common conventional steam methane reforming method (11.89 kg CO_2eq /kgH_2).

Also, the percentage results of GWP in Fig. 4 for the different stages in the life cycle (construction, operation, and waste treatment) confirmed the greater contribution of the used electricity supply to the GWP values in the four scenarios. It can be also noticed that the contribution of the electrolyzer system to the total GWP impact increased from the grid mix scenario (lower than 1%) to the PV solar (lower than

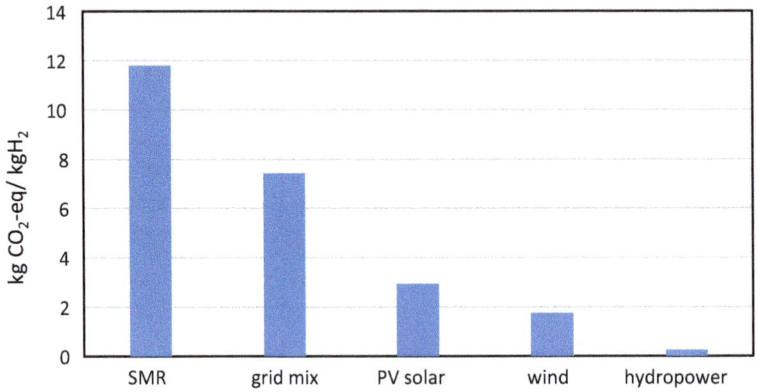

Fig. 3 GWP (kg CO_2 eq per kg H_2) for the total LCA in the different four scenarios compared to SMR method

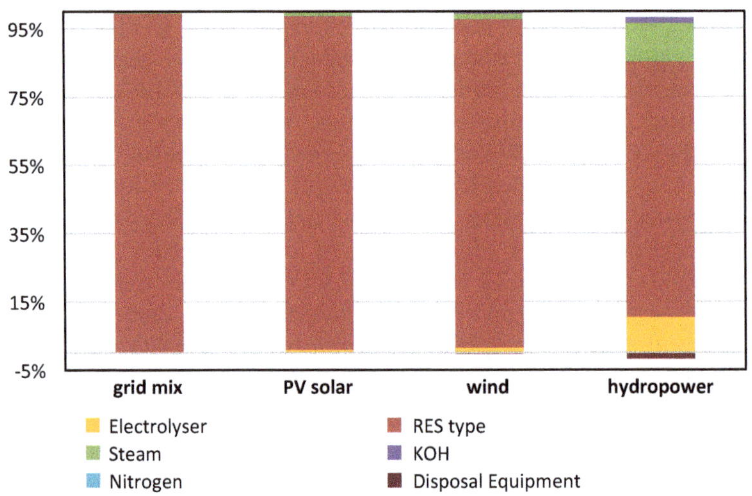

Fig. 4 GWP in percentages for the total LCA in the different scenarios

2%) to the wind scenario (lower than 4%) to the hydropower scenario (around 16%). The impacts of the electrolyzer system are mainly due to its construction requirements and components' materials such as steel, copper, and nickel that counted for 20 years lifetime of the project. Also, as the hydropower electricity has the minimum value of GWP compared to other three RES, the percentage contribution of the electrolyzer system appears to be higher than the other three scenarios. Also, the influence of the disposal equipment phase (waste treatment) appeared more in the hydropower scenario with around −2% contribution of GWP decrease. In the disposal phase of some materials in the electrolyzer system, there is an additional low electricity supply required for the recycling process that doesn't allow high recovery levels in GWP.

Therefore, the development of the electrolyzer technology, especially its conventional energy efficiency that usually falls within the range of 54% to 68%, can lead to lowering the consumption of needed electricity and some critical materials, thus reducing the environmental impacts of the hydrogen production process. In this, ongoing research is being conducted to advance electrolyzer technology, with the aim of mitigating its drawbacks. This includes improvements in areas such as current density, operating pressure, electro-catalysts, lifetime, and energy efficiency [43, 44].

It should be also mentioned that parameters inputs associated with electrolysis hydrogen production, including the operation (electricity consumption) and construction, may change during the lifetime of the project causing a change in GWP. Therefore, an optimization of the operating conditions, which can be also related to the electricity generation of RES, significantly contributes to the advancement of electrolysis hydrogen production linked to RES using high current density, zero-gap design, less membrane thickness, and suitable operating pressure [45]. The optimization of electrolyzer performance necessitates a meticulous focus on the design of its components such as the electrodes, as well as the electricity generation conditions associated with the location (in Portugal for this study). In addition, maintaining stable power performance is crucial to avoid the dips in energy supply that can occur at certain times of the day due to environmental factors. For instance, the energy output from photovoltaic (PV) solar systems can fluctuate based on the number of sunny hours in a day. Therefore, it's important to have strategies in place to manage these variations and ensure a consistent energy feed. This could involve using energy storage systems or integrating with other renewable energy sources to balance the supply [46].

However, the GWP results found in the current study are not far from related literature. Ghandehariun and Kumar [47] documented a GWP of approximately 0.68 kg CO_2 eq./kg H_2 for hydrogen production based on wind energy. Additionally, Sadeghi et al. [51] identified a GWP value of 3.08 kg CO_2-eq./kg H_2 for a hydrogen production system powered by PV solar energy. Conversely, Zhang et al. [48] observed GWP values ranging from 8.67 to 4.33 kg CO_2 eq/kg H_2 (varying with the system's lifetime from 30 to 60 years) for an electrolysis system utilizing PV power. Generally, prior research in this domain has primarily focused on the overall LCA of hydrogen production [19, 49, 50]. However, to the best of the authors' knowledge, there has been no comprehensive study concentrating on various RES scenarios, as outlined in this study, in comparison to an updated electricity grid mix scenario, particularly for the Portugal case study. Nevertheless, environmental impact assessments of hydrogen production have exhibited considerable variation across existing research, attributed to factors such as system capacity, energy source, geographical conditions, and electrolysis technology materials.

Even though the hydropower scenario appeared to cause the minimum GWP impact for producing GH_2, solar and wind may be favored in some cases due to the numerous benefits of incorporating solar and wind into an H_2 production system compared to other RES options. While hydropower is often better suited for providing a stable base load to the grid without requiring the balancing of input sources

[51]. In this context, solar energy stands out for its ubiquitous availability and infinite nature, making it the foremost recommended clean energy source globally [52].

5 Conclusion

The current study has analyzed the life cycle of GH_2 production by Alkaline electrolysis for different renewable energy sources in Portugal. Four electricity supply technology scenarios for the electrolyzer system were investigated, namely, hydropower, wind turbine, PV solar, and the electricity grid mix in Portugal. LCA results demonstrated that electricity supply dominates the highest contribution percentage of GWP values among the different stages in the life cycle (construction, operation, and waste treatment). Also, hydrogen produced from the hydropower scenario has the lowest GWP impact of 0.26 kg CO_2eq /kgH_2, followed by 1.76 for the wind turbine scenario, followed by 2.94 for the PV solar scenario, and 7.4 kg CO_2eq / kgH_2 for the grid mix scenario. The use of the hydropower scenario presented a GWP value lower by −97% compared to the Grid mix scenario. While the wind turbine and solar scenarios were lower by −76% and −76% compared to the Grid mix scenario, respectively. Also, the Grid mix scenario exhibited a good GWP impact of 7.4 kg CO_2-eq /kgH_2 compared to the conventional SMR method for hydrogen production, due to the high percentages of multi-RES contributing to Portuguese grid electricity. In general, all the GWP values obtained are lower than the common conventional steam methane reforming method (11.89 kg CO_2eq / kgH_2). The latter highlights the significant advantages of implementing these four scenarios and fostering collaboration among them to enhance the environmental performance of hydrogen production systems.

References

1. "Our World in Data. Renewable Energy," *https://ourworldindata.org/renewable-energy*, 2023.
2. R. Daiyan, I. Macgill, and R. AmalOpportunities and Challenges for Renewable Power-to-X," *ACS Energy Lett.*, vol. 5, no. 12, pp. 3843–3847, 2020, doi: https://doi.org/10.1021/acsenergylett.0c02249.
3. K. Adeli, M. Nachtane, A. Faik, D. Saifaoui, and A. Boulezhar, "How Green Hydrogen and Ammonia Are Revolutionizing the Future of Energy Production: A Comprehensive Review of the Latest Developments and Future Prospects," *Appl. Sci.*, vol. 13, no. 15, 2023, doi: https://doi.org/10.3390/app13158711.
4. G. Chisholm, T. Zhao, and L. Cronin, *Hydrogen from water electrolysis*. Elsevier Inc., 2022. doi: https://doi.org/10.1016/B978-0-12-824510-1.00015-5.
5. W. Ajeeb, D. Melo, R. Costa, and P. Baptista, "Life cycle analysis of hydrotreated vegetable oils production based on green hydrogen and used cooking oils," Fuel, vol. 390, no. February, p. 134–749, 2025, https://doi.org/10.1016/j.fuel.2025.134749.
6. L. Wang *et al.*, "Power-to-fuels via solid-oxide electrolyzer: Operating window and techno-economics," *Renew. Sustain. Energy Rev.*, vol. 110, no. April, pp. 174–187, 2019, doi: https://doi.org/10.1016/j.rser.2019.04.071.

7. T. Yusaf *et al.*, "Sustainable hydrogen energy in aviation – A narrative review," *Int. J. Hydrogen Energy*, vol. 52, pp. 1026–1045, 2024, doi: https://doi.org/10.1016/j.ijhydene.2023.02.086.
8. J. M. Thomas, P. P. Edwards, P. J. Dobson, and G. P. Owen, "Decarbonising energy: The developing international activity in hydrogen technologies and fuel cells," *J. Energy Chem.*, vol. 51, pp. 405–415, 2020, doi: https://doi.org/10.1016/j.jechem.2020.03.087.
9. L. Phan-Van, V. N. Dinh, R. Felici, and T. N. Duc, "New models for feasibility assessment and electrolyser optimal sizing of hydrogen production from dedicated wind farms and solar photovoltaic farms, and case studies for Scotland and Vietnam," *Energy Convers. Manag.*, vol. 295, no. September, p. 117597, 2023, doi: https://doi.org/10.1016/j.enconman.2023.117597.
10. R. Assunçao, W. Ajeeb, F. Eckl, D. Gomes R. Neto, "Decentralized hydrogen-oxygen co-production via electrolysis for large hospitals with integrated hydrogen refueling station," *Int. J. HydrogenEnergy.,* vol. 103, pp. 87–98, 2025, doi: https://doi.org/10.1016/j.ijhydene.2025.01.169.
11. M. Khalil and I. Dincer, "Investigation of a community-based clean energy system holistically with renewable and hydrogen energy options for better sustainable development," *J. Clean. Prod.*, vol. 440, no. December 2023, p. 140895, 2024, doi: 10.1016/j.jclepro.2024.140895.
12. Q. Hassan, A. Z. Sameen, H. M. Salman, and M. Jaszczur, "Large-scale green hydrogen production via alkaline water electrolysis using solar and wind energy," *Int. J. Hydrogen Energy*, vol. 48, no. 88, pp. 34299–34315, 2023, doi: https://doi.org/10.1016/j.ijhydene.2023.05.126.
13. Comissão Europeia, "Plano Nacional Energia E Clima 2021-2030 (Pnec 2030)," 2018.
14. F. Suleman, I. Dincer, and M. Agelin-Chaab, "Environmental impact assessment and comparison of some hydrogen production options," Int. J. Hydrogen Energy, vol. 40, no. 21, pp. 6976–6987, 2015, doi: https://doi.org/10.1016/j.ijhydene.2015.03.123.
15. K. Farhana, A. Shadate Faisal Mahamude, and K. Kadirgama, "Comparing hydrogen fuel cost of production from various sources - a competitive analysis," *Energy Convers. Manag.*, vol. 302, no. January, p. 118088, 2024, doi: https://doi.org/10.1016/j.enconman.2024.118088.
16. P. L. Spath and M. K. Mann, "Life Cycle Assessment of Hydrogen Production via Natural Gas Steam Reforming," Golden, Colorado (US), 2000. [Online]. Available: https://www.nrel.gov/docs/fy01osti/27637.pdf
17. A. Elgowainy, J. Han, and H. Zhu, "Updates to Parameters of Hydrogen Production," pp. 2–5, 2013, [Online]. Available: https://greet.es.anl.gov/files/h2-13
18. J. C. Koj, C. Wulf, A. Schreiber, and P. Zapp, "Site-dependent environmental impacts of industrial hydrogen production by alkalinewater electrolysis," *Energies*, vol. 10, no. 7, 2017, doi: https://doi.org/10.3390/en10070860.
19. X. Zhang, C. Bauer, C. L. Mutel, and K. Volkart, "Life Cycle Assessment of Power-to-Gas: Approaches, system variations and their environmental implications," *Appl. Energy*, vol. 190, pp. 326–338, 2017, doi: https://doi.org/10.1016/j.apenergy.2016.12.098.
20. Zulfhazli, A. R. Keeley, S. Takeda, and S. Managi, "A systematic review of the techno-economic assessment of various hydrogen production methods of power generation," *Front. Sustain.*, vol. 3, 2022, doi: https://doi.org/10.3389/frsus.2022.943145.
21. "International Organization for Standardization (ISO). Environmental management e life cycle assessment e principles and framework," 2006. https://www.iso.org/obp/ui/#iso:std:iso:14040:ed-2:v1:en
22. A. Lozanovski, O. Schuller, M. Faltenbacher, M. Fischer, K. Sedlbauer, and Fraunhofer-Institut für Bauphysik, *Guidance document for performing LCA on hydrogen production systems*, vol. 9. 2013.
23. European Comission, "Recommendations for Life Cycle Impact Assessment in the European context - Based on existing environmental impact assessment models and factors," 2011.
24. S. Seetharaman, R. Balaji, K. Ramya, K. S. Dhathathreyan, and M. Velan, "Graphene oxide modified non-noble metal electrode for alkaline anion exchange membrane water electrolyzers," *Int. J. Hydrogen Energy*, vol. 38, no. 35, pp. 14934–14942, 2013, doi: https://doi.org/10.1016/j.ijhydene.2013.09.033.
25. I. Abe, "Alkaline Water Electrolysis. In Energy Carriers and Conversion Systems," *UNESCO–Encyclopedia Life Support Syst. Chiba*, vol. I, 2009.

26. S. Shiva Kumar, S. U. B. Ramakrishna, S. V. Krishna, K. Srilatha, B. R. Devi, and V. Himabindu, "Synthesis of titanium (IV) oxide composite membrane for hydrogen production through alkaline water electrolysis," *South African J. Chem. Eng.*, vol. 25, pp. 54–61, 2018, doi: https://doi.org/10.1016/j.sajce.2017.12.004.
27. *W. Ajeeb, R. Costa, and P. Baptista*, "Life cycle assessment of green hydrogen production through electrolysis : A literature review," Sustain. Energy Technol. Assessments, vol. 69, pp.103923, 2024, doi: https://doi.org/10.1016/j.seta.2024.103923.
28. W. Ajeeb, P. Baptista, and R. Costa, "Life cycle analysis of hydrogen production by different alkaline electrolyser technologies sourced with renewable energy," Energy Convers. Manag., vol. 316, no. 14, p. 118840, 2024, doi: https://doi.org/10.1016/j.enconman.2024.118840.
29. J. Brauns and T. Turek, "Model-Based Analysis and Optimization of Pressurized Alkaline Water Electrolysis Powered by Renewable Energy," *J. Electrochem. Soc.*, vol. 170, no. 6, p. 064510, 2023, doi: https://doi.org/10.1149/1945-7111/acd9f1.
30. Nel Hydrogen, "The World's Most Efficient and Reliable Electrolysers," *Nel Hydrog.*, p. 16, 2020, [Online]. Available: https://nelhydrogen.com/wp-content/uploads/2020/03/Electrolysers-Brochure-Rev-C.pdf
31. L. Correas, "Improvements to Integrate High Pressure Alkaline Electrolysers for Electricity / H2 production from Renewable Energies to Balance the Grid," no. 1, pp. 1–12, 2014.
32. W. Ajeeb and S. M. S. Murshed, "Pool boiling heat transfer characteristics of new and recycled alumina Nanofluids," Nanomaterials, vol. 13, p. 1040, 2023, https://doi.org/10.3390/nano13061040.
33. W. Ajeeb and S. M. S. Murshed, "Pool boiling heat transfer characteristics of SiO 2 and BN Nanoparticles dispersed mono and hybrid Nanofluids," Nanomaterials, vol. 11, pp. 1–13, 2023.
34. W. Ajeeb and S. M. S. Murshed, "Comparisons of numerical and experimental investigations of the thermal performance of Al2O3 and TiO2 Nanofluids in a compact plate heat exchanger," Nanomaterials, vol. 12, no. 20, p. 3634, 2022, https://doi.org/10.3390/nano12203634.
35. W. Ajeeb, D. Roque, and S. M. S. Murshed, "Thermal-hydraulic performance of Al2O3 Nanofluids under laminar flow in a Minitube," Int. J. Energy a Clean Environ., vol. 23, no. 5, pp. 39–59, 2022, https://doi.org/10.1615/InterJEnerCleanEnv.2022041476.
36. W. Ajeeb and S. M. S. Murshed, "Characterization of Thermophysical and electrical properties of SiC and BN Nanofluids," Energies, vol. 16, no. 9, 2023, https://doi.org/10.3390/en16093768.
37. "European Communities: Wind Turbine Grid Connection and Interaction," 2001. [Online]. Available: https://energy.ec.europa.eu/publications/wind-turbine-grid-connection-and-interaction_en
38. Eurelectric, "Composition of Portugal's electricity generation (2023)," 2023. https://electricity-data.eurelectric.org/electricity-generation/electricity-generation-by-fuel.html
39. Eurelectric, "Composition of Spain's electricity generation (2023)," 2023. https://electricity-data.eurelectric.org/electricity-generation/electricity-generation-by-fuel.html
40. A. Lotrič, M. Sekavčnik, I. Kuštrin, and M. Mori, "Life-cycle assessment of hydrogen technologies with the focus on EU critical raw materials and end-of-life strategies," *Int. J. Hydrogen Energy*, vol. 46, no. 16, pp. 10143–10160, 2021, doi: https://doi.org/10.1016/j.ijhydene.2020.06.190.
41. L. Schleisner, "Life cycle assessment of a wind farm and related externalities," *Renew. Energy*, vol. 20, no. 3, pp. 279–288, 2000, doi: https://doi.org/10.1016/S0960-1481(99)00123-8.
42. D. H. N. C. Reichel, A. Müller, L. Friedrich, S. Herceg, M. Mittag, A. Protti, "CO2 Emissions of Silicon Photovoltaic Modules – Impact of Module Design and Production Location," *8th World Conf. Photovolt. Energy Convers.*, no. September, pp. 1617–1619, 2022.
43. A. Pozio, F. Bozza, G. Nigliaccio, M. Platter, and G. Monteleone, "Development perspectives on low-temperature electrolysis," *Energia, Ambient. e Innov.*, no. July, pp. 66–71, 2021, doi: https://doi.org/10.12910/EAI2021-014.

44. S. Anwar, F. Khan, Y. Zhang, and A. Djire, "Recent development in electrocatalysts for hydrogen production through water electrolysis," *Int. J. Hydrogen Energy*, vol. 46, no. 63, pp. 32284–32317, 2021, doi: https://doi.org/10.1016/j.ijhydene.2021.06.191.
45. M. T. de Groot, J. Kraakman, and R. L. Garcia Barros, "Optimal operating parameters for advanced alkaline water electrolysis," *Int. J. Hydrogen Energy*, vol. 47, no. 82, pp. 34773–34783, 2022, doi: https://doi.org/10.1016/j.ijhydene.2022.08.075.
46. R. Li *et al.*, "Capacity optimization of a wind-photovoltaic-electrolysis-battery (WPEB) hybrid energy system for power and hydrogen generation," *Int. J. Hydrogen Energy*, no. xxxx, 2023, doi: https://doi.org/10.1016/j.ijhydene.2023.08.029.
47. S. Ghandehariun and A. Kumar, "Life cycle assessment of wind-based hydrogen production in Western Canada," *Int. J. Hydrogen Energy*, vol. 41, no. 22, pp. 9696–9704, 2016, doi: https://doi.org/10.1016/j.ijhydene.2016.04.077.
48. J. Zhang, B. Ling, Y. He, Y. Zhu, and Z. Wang, "Life cycle assessment of three types of hydrogen production methods using solar energy," *Int. J. Hydrogen Energy*, vol. 47, no. 30, pp. 14158–14168, 2022, doi: https://doi.org/10.1016/j.ijhydene.2022.02.150.
49. N. Gerloff, "Comparative Life-Cycle-Assessment analysis of three major water electrolysis technologies while applying various energy scenarios for a greener hydrogen production," *J. Energy Storage*, vol. 43, no. May, p. 102759, 2021, doi: https://doi.org/10.1016/j.est.2021.102759.
50. M. S. Henriksen *et al.*, "Tradeoffs in life cycle water use and greenhouse gas emissions of hydrogen production pathways," *Int. J. Hydrogen Energy*, 2023, doi: https://doi.org/10.1016/j.ijhydene.2023.08.079.
51. G. Gahleitner, "Hydrogen from renewable electricity: An international review of power-to-gas pilot plants for stationary applications," *Int. J. Hydrogen Energy*, vol. 38, no. 5, pp. 2039–2061, 2013, doi: https://doi.org/10.1016/j.ijhydene.2012.12.010.
52. Q. Hassan, M. Jaszczur, J. Teneta, M. K. Abbas, A. Hasan, and A. K. Al-Jiboory, "Experimental investigation for the estimation of the intensity of solar irradiance on oblique surfaces by means of various models," *Energy Harvest. Syst.*, vol. 9, pp. 227–237, 2022, doi: https://doi.org/10.1515/ehs-2021-0087.

Optimizing Disc Screens for Efficient Waste Separation

Yangzhen Gao, Liang Han, Jian Zhang, and Yingze Lin

Abstract With rapid urbanization, effectively separating household and construction waste is vital for environmental protection and resource recovery. This study aims to optimize the performance of disc screen equipment in separating these waste types. By adjusting the inclination angle of the disc screen and the shaft speed of the discs, the optimal parameter settings were determined to maximize separation efficiency and accuracy. The triangular disc screen equipment XS-S-1560, featuring adjustable inclination angle and shaft speed, was utilized for the experiments. Household and construction waste from the Yangzhou region of China were used as experimental materials. The results demonstrated that, for household waste, an inclination angle of 0° and a shaft speed of 75 rpm achieved a screening efficiency of 97.507%. For construction waste, an inclination angle of 14° and a shaft speed of 80 rpm resulted in a screening efficiency of 94.949%. These findings provide scientific evidence for the application of disc screen equipment in material recycling facilities, contributing to enhanced economic and environmental sustainability. The developed regression models provide reliable predictions for optimizing disc screen operations, ensuring higher purity and market value of recycled materials. This research underscores the importance of tailored optimization for different waste compositions and highlights the potential for further advancements in solid waste treatment technologies.

Keywords Disc screen optimization · Municipal solid waste · Screening efficiency · Household waste · Construction waste

Y. Gao · L. Han (✉) · Y. Lin
School of Mechanical Engineering, Southeast University, Nanjing, China
e-mail: melhan@seu.edu.cn

J. Zhang
Mizuda Xinzhizao Huzhou Technology Co., Ltd., Huzhou, China

Z. Peng (ed.), *Technological Advancements and Future Directions in Green Energy*, Green Energy and Technology,
https://doi.org/10.1007/978-3-031-83203-1_5

1 Introduction

With the acceleration of global urbanization, the production of municipal solid waste, including household waste and construction waste, is increasing, posing significant challenges to land use, environmental health, and resource conservation [1]. Municipal solid waste includes waste generated from residential, commercial, and public services. Effective management and resource utilization have become major challenges for sustainable urban development worldwide [2].

As the world's largest developing country, China has experienced a rapid increase in municipal solid waste due to economic growth and continuous population growth, making waste management particularly urgent. Statistics show that China generates billions of tons of solid waste annually, including a large amount of household and construction waste, which increases the difficulty of disposal [3]. To address this issue, the Chinese government has introduced policies to promote waste segregation, strengthen waste management, and increase investment in recycling and resource utilization to build a comprehensive network of materials recovery facility (MRF) [4–6]. MRF in various cities effectively extracts recyclable materials from municipal solid waste through processes such as collection, sorting, crushing, and washing, using various types of waste sorting equipment, thereby reducing the environmental risks associated with landfill or incineration [7]. In China, the design and operation of MRF are advancing towards automation and intelligence to meet the growing demand for waste treatment [8].

Disc screens are typically used as front-end screening devices in MRF to separate recyclable municipal solid waste from non-recyclable materials. The performance of the disc screen is crucial in determining the purity and market value of the final recycled products [9]. By adjusting parameters such as the spacing between shafts, the shape of the discs, and the rotational speed, disc screens can achieve precise screening of materials of different particle sizes, thereby significantly enhancing the efficiency of subsequent recycling and treatment processes. However, optimizing the performance of disc screens faces many challenges, including the diversity of material physical properties, the adaptability of screen specifications, and the complexity of equipment structural parameters. These factors collectively limit efficiency and accuracy in practical operations. In China, considering the significant regional and intercity differences in municipal solid waste composition, designing and optimizing disc screens tailored to the unique characteristics of waste in different areas is particularly urgent.

In recent years, scholars domestically and internationally have extensively studied the use of disc screens in waste treatment, mineral processing, agriculture, and biomass energy conversion. Arina et al. compared the performance of star screens and trommel screens in processing municipal solid waste, revealing the significant impact of different screening equipment structures on screening outcomes [10]. Rezaee et al. used response surface methodology to evaluate the key factors influencing disc screens in waste paper sorting, optimizing screening parameters to significantly improve classification efficiency, and demonstrating the advantages of disc screens in improving the quality of recycled materials [11]. Zhao Lala et al.

designed a cross-screening experiment for wet fine coal and verified the effect of screen structure on screening efficiency through DEM simulation [12]. Huber et al. designed a mobile star screen to improve the quality of wood chips from forest residues, providing a more efficient pretreatment method for biomass energy conversion [13]. Additionally, Bulgakov et al. developed a new spiral potato cleaner, effectively removing soil clods and impurities, proving the importance of optimized screening devices in enhancing resource recovery efficiency in specific agricultural waste treatments [14]. Lv Jinqing et al. designed a grading device for potato seed tuber cutters and experimentally validated the feasibility of improving grading efficiency by adjusting equipment structural parameters, providing valuable experience for screening specific materials [15]. Silva et al. improved the screening efficiency of roller screens for iron ore pellets through statistical analysis and discrete element simulation [16].

Despite these advancements, in-depth discussions on improving screening efficiency and optimizing parameters for specific types of municipal solid waste, especially construction waste and household waste, remain limited. Therefore, building on the above research findings, this study focuses on the efficient separation of construction waste and household waste. It aims to optimize the inclination angle of the screen surface and the rotational speed of the discs to enhance screening efficiency, reduce impurity content, and improve the economic and environmental sustainability of recycling processes. This research provides theoretical support and practical guidance for the actual operation of MRF, promoting more efficient and environmentally friendly waste resource utilization.

2 Materials and Methods

2.1 Disc Screen Equipment

The triangular disc screen equipment XS-S-1560 used in this study (see Fig. 1) features adjustable inclination angle and shaft speed. The dimensions of the screen body, aperture sizes, and other technical specifications are designed to replicate the conditions of a real MRF (Table 1).

2.2 Experimental Materials

Construction waste and household waste from the Yangzhou region in China were collected as experimental materials. The compositions of these materials are presented in Tables 2 and 3. The materials were categorized into three size grades: Fines (<55 mm), Acceptable (55 ~ 65 mm), and Oversize (>65 mm).

Fig. 1 The XS-S-1560 disc screen

Table 1 Main technical parameters of the triangular disc screen XS-S-1560 equipment

Parameter	Value
Machine size (L × W × H)	2552 mm × 1532 mm × 1050 mm
Inclination angle	0 ~ 30°
Number of shafts	11
Number of metal disc	17
Number of rubber disc	15
Shaft distance	222 mm
Distance between discs	58.5 mm
Shaft speed	0 ~ 150 rpm
Max. throughput capacity	40 m^3/h

Table 2 Composition of construction waste

Material	Components	Proportion
Concrete and sand	Concrete, sand	55%
Wood	Wood planks	13%
Paper products	Cardboard, cartons	12%
Plastic products	Plastic films, foam	9%
Metal	Rebars, aluminum alloys	6%
Others	Glass, others	5%

Table 3 Composition of household waste

Material	Components	Proportion
Plastic products	Bottles, plastic packaging	38%
Paper products	Wood planks, cigarette packs, milk cartons	27%
Organic materials	Leaves	3%
Metal	Cans	13%
Stones	Stones	11%
Others	Glass, others	8%

2.3 *Force Analysis on the Disc Screen*

Materials are conveyed to the disc screen equipment via a conveyor belt and move forward with the rotation of the disc screen. The forces acting on the material on two adjacent disc screens are shown (see Fig. 2).

In the figure:

$G = mg$ represents the gravity of the material,
N_1 and N_2 are the support forces exerted by disc 1 and disc 2 on the material, respectively,
F_1 and F_2 are the friction forces from the discs 1 and 2 on the material,
α is the angle between the support forces N_1 and N_2,
β is the angle between the gravitational force and the normal at the contact point C_2on disc 2,
θ is the inclination angle of the disc plane,
n is the rotational speed of the discs.

The material on the disc screen is primarily subjected to five forces: its gravitational force G, the support forces from the two adjacent discs N_1 and N_2, and the two friction forces F_1 and F_2. For the material to move continuously and smoothly forward on the disc screen, effective screening, the resultant moment at the contact point C_2 on disc 2 must satisfy $\Sigma M_{c_2} \geq 0$ [15, 17], i.e.,

$$\Sigma M_{c_2} = N_1 R \sin\left(180^\circ \ \alpha\right) - F_1 R\left[1 + \cos\left(180^\circ \ \alpha\right)\right] - GR\sin\beta \geq 0 \tag{1}$$

2.4 *Disc Screen Screening Efficiency*

The calculation formula for disc screen screening efficiency is

$$\eta = \frac{m_1}{m_2} \times 100\% \tag{2}$$

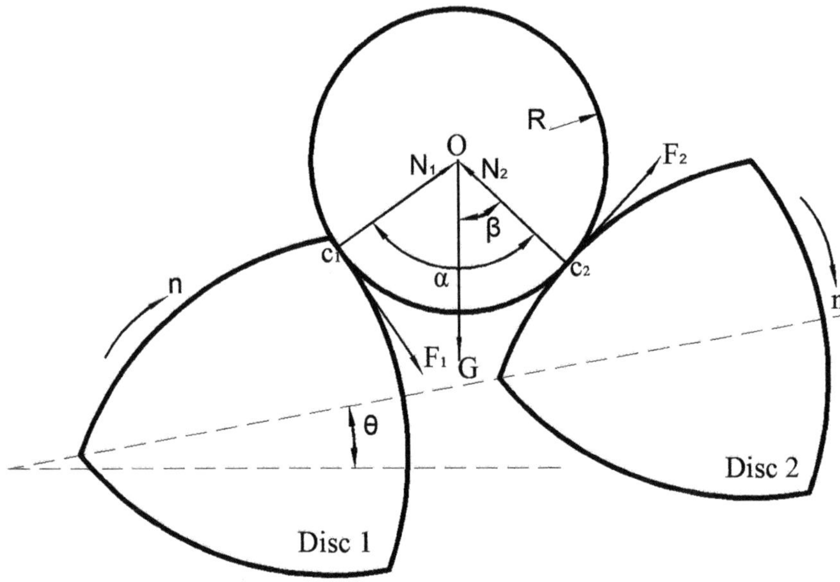

Fig. 2 Force analysis of material on a disc screen

In the equation:

η represents the screening efficiency (%),

m_1 represents the mass of particles smaller than the sieve aperture size in the material below the disc screen surface (kg),

m_2 represents the mass of particles smaller than the sieve aperture size in the total screened material (kg).

3 Experiments and Analysis

3.1 Experimental Design

The screening process of household waste and construction waste on the disc screen equipment is influenced by factors such as the screen surface's inclination angle and the discs' shaft speed. To develop a predictive model for screening efficiency, a series of experiments were conducted on household and construction waste. These experiments varied the inclination angle and shaft speed of the disc screen while maintaining a constant input quantity of material. The goal was to assess the impact of different parameters on separation efficiency.

3.2 Experimental Data Analysis

In this study, the Central Composite Design (CCD) method in Design Expert 13.0.5 was used for regression analysis to develop a mathematical model relating the response value to each factor. Analysis of variance (ANOVA) was employed to determine the significance of each factor and the validity of the model. The effects of the disc screen surface inclination angle and the screening speed on the screening efficiency of different types of waste were investigated.

The experimental design and results for household waste are presented in Table 4, and those for construction waste are shown in Table 5. The Analysis module in Design Expert was used for simulation fitting, variance analysis, significance testing of equations, and significance testing of coefficients. Based on the experimental results, a Response Surface Methodology (RSM) model was developed, and 3D response surface plots and contour plots illustrating the relationship between screening efficiency and influencing factors for household and construction waste were generated.

Table 4 Experimental design and results of daily life waste

	Factor		
Run	Inclination angle (°)	Shaft speed (rpm)	Screening efficiency (%)
1	0	45	90
2	0	67	96
3	0	84	99
4	0	90	95
5	7.5	45	89
6	7.5	67	95
7	7.5	84	93
8	7.5	90	92
9	11	45	86
10	11	67	92
11	11	84	96
12	11	90	95
13	15	45	82
14	15	67	94
15	15	84	92

Table 5 Experimental design and results of construction waste

Run	Factor		Screening efficiency (%)
	Inclination angle (°)	Shaft speed (rpm)	
1	0	45	90
2	0	78	92
3	0	90	86
4	7.5	23	74
5	7.5	45	79
6	7.5	78	98
7	7.5	90	92
8	11	45	86
9	11	78	92
10	11	90	89
11	14	45	86
12	14	78	97
13	14	90	93

4 Results and Discussion

4.1 *Household Waste Screening*

Using Design Expert, the second-order polynomial regression equation relating the response value of Screening Efficiency (Y_1) to the factors Inclination Angle (A_1) and Shaft Speed (B_1) for household waste screening was derived as follows:

$$Y_1 = 52.1289 - 0.698567A_1 + 1.22773B_1 + 0.00590915A_1B_1 - 0.00210769A_1^2 - 0.00830426B_1^2 \quad (3)$$

The regression equation reveals that factor A_1 is negatively correlated with the dependent variable Y_1, while factor B_1 is positively correlated with Y_1. The regression coefficient R^2 is 0.8657, indicating a small deviation between the predicted and actual values. This suggests that 86.57% of the experimental results can be explained and predicted by this model, demonstrating that the established regression model effectively reflects the functional relationship between A_1, B_1, and the response value Y_1.

The ANOVA for the screening efficiency of household waste is presented in Table 6. The model's *P*-value <0.01 indicates a strong regression effect and high significance. The regression coefficient signifies that the model has high reliability and accurately reflects the actual values. Additionally, with Adeq Precision>4, the model demonstrates a good fit between the predicted and experimental values.

Figure 3 illustrates the impact of the interaction between the disc screen inclination angle and screening speed on the screening efficiency of household waste. The screening efficiency decreases linearly as the inclination angle increases. With an increase in screening speed, the efficiency first rises rapidly and then declines slowly. The 3D response surface plot is relatively flat, and the contour plot is sparse.

Table 6 Variance analysis for the established regression model

Source	Sum of squares	DOF	Mean square	F-value	P-value	Comments
Model	224.73	5	44.95	11.60	0.0010	Significant
A_1-Inclination angle	44.8	1	44.80	11.56	0.0079	
B_1-Shaft speed	99.35	1	99.35	25.64	0.0007	
A_1B_1	4.37	1	4.37	1.13	0.3158	
A_1^2	0.0395	1	0.0395	0.0102	0.9218	
B_1^2	43.48	1	43.48	11.22	0.0085	
Residual	34.87	9	3.87			
Cor total	259.60	14				

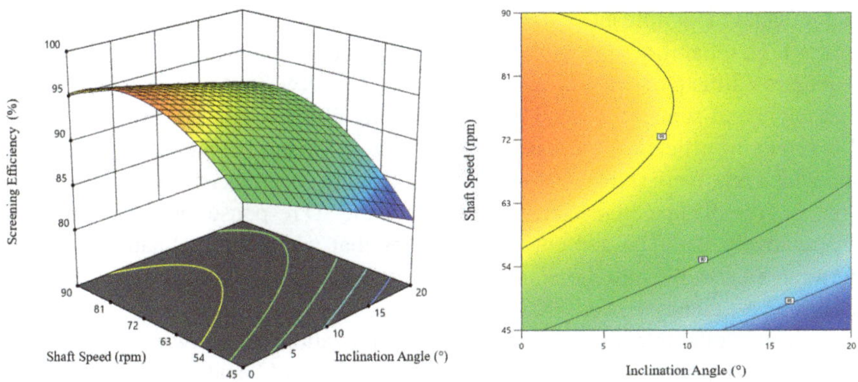

(a) 3D response curve (b) Contour plot

Fig. 3 Interaction between inclination angle and shaft speed on the screening efficiency

The contour plot shows that when the Inclination Angle is between 0° and 9° and the Shaft speed is between 57 and 90 rpm, the predicted screening efficiency is high, exceeding 95%.

Using the effective regression model for optimization, the software predicted the optimal parameters to be an Inclination Angle of 0° and a Shaft Speed of 73.927 rpm, with a predicted Screening Efficiency of 97.507%. Considering experimental conditions, operability, and feasibility, the Inclination Angle can be set to 0° and the Shaft Speed to 75 rpm in practical experiments.

4.2 *Construction Waste Screening*

Similarly, using Design Expert, a second-order polynomial regression equation was derived for the response value Screening Efficiency (Y_2) about the factors Inclination Angle (A_2) and Shaft Speed (B_2) for construction waste screening.

$$Y_2 = 57.6677 - 1.27653A_2 + 1.05625B_2 + 0.0155289A_2B_2 + 0.0226078A_2^2 - 0.00799555B_2^2 \tag{4}$$

The regression equation reveals that factor A_2 is negatively correlated with the dependent variable Y_2, while factor B_2 is positively correlated with Y_2. The regression coefficient R^2 is 0.7932, indicating a small deviation between the predicted and actual values, and that 79.32% of the experimental results can be explained and predicted by this model. This demonstrates that the established model effectively reflects the functional relationship between A_2, B_2, and the response value Y_2.

ANOVA for the fitted model of response value Y_2, including the significance of each factor, is shown in Table 7. The model's P-value <0.05 indicates a good regression effect and significant correlation. With Adeq Precision >4, the model demonstrates high reliability and a good fit with the experimental data.

Figure 4 illustrates the impact of the interaction between the disc screen inclination angle and screening speed on the screening efficiency of construction waste. The screening efficiency decreases linearly with an increase in the inclination angle, showing a relatively stable decline. As the screening speed increases, the efficiency first rises rapidly and then declines slowly, with the 3D response surface plot appearing relatively flat. The contour plot indicates that when the Inclination Angle is between 8° and 14° and the Shaft Speed is between 60 and 90 rpm, the predicted screening efficiency exceeds 90%.

Using the effective regression model for optimization, the software predicted the optimal parameters to be an Inclination Angle of 14° and a Shaft Speed of 79.648 rpm, with a predicted screening efficiency of 94.949% for construction waste. Considering experimental conditions, operability, and feasibility, the Inclination Angle can be set to 14° and the Shaft Speed to 80 rpm in practical experiments.

Table 7 Variance analysis for the established regression model

Source	Sum of squares	DOF	Mean square	F-value	P-value	Comments
Model	428.58	5	85.72	5.37	0.0240	Significant
A-Inclination angle	1.40	1	1.40	0.0877	0.7758	
B-Shaft speed	297.19	1	297.19	18.62	0.0035	
AB	28.62	1	28.62	1.79	0.2224	
A^2	2.75	1	2.75	0.1724	0.6904	
B^2	104.03	1	104.03	6.52	0.0379	
Residual	111.72	7	15.96			
Cor total	540.31	12				

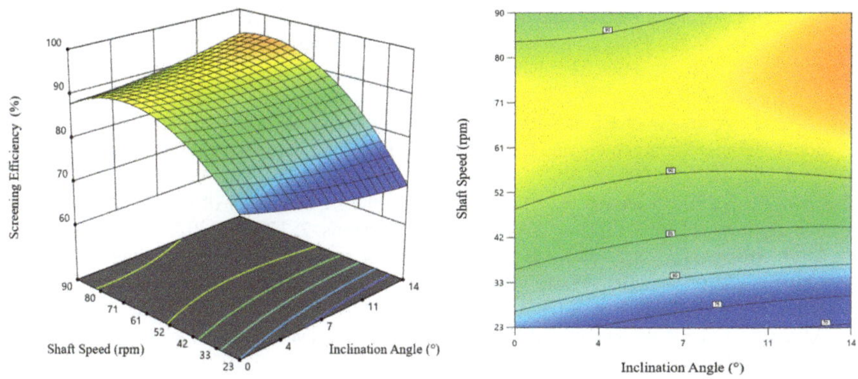

(a) 3D response curve (b) Contour plot

Fig. 4 Interaction between inclination angle and shaft speed on the screening efficiency

5 Conclusion

The optimization of disc screen equipment for the separation of household waste and construction waste has been thoroughly investigated in this study. By adjusting the inclination angle of the disc screen and the shaft speed of the discs, the screening efficiency and accuracy were significantly improved. The experimental results demonstrated that an inclination angle of 0° and a shaft speed of 75 rpm is optimal for household waste, achieving a screening efficiency of 97.507%. For construction waste, the optimal parameters were identified as an inclination angle of 14° and a shaft speed of 80 rpm, resulting in a screening efficiency of 94.949%.

These findings provide valuable insights into the practical application of disc screen equipment in MRF. The improved screening efficiency contributes to the reduction of impurity content and enhances the economic and environmental sustainability of recycling processes. The regression models developed provide reliable predictions for optimizing disc screen operations, ensuring higher purity and market value of recycled materials. This research underscores the importance of tailored optimization for different waste compositions and highlights the potential for further advancements in solid waste treatment technologies.

References

1. Zambrano-Monserrate, M.A., Ruano, M.A., Ormeño-Candelario, V.: Determinants of municipal solid waste: a global analysis by countries' income level. Environ Sci Pollut R. 28, 62421–62430 (2021).
2. Khan, S., Anjum, R., Raza, S.T., Ahmed Bazai, N., Ihtisham, M.: Technologies for municipal solid waste management: Current status, challenges, and future perspectives. Chemosphere. 288, 132403 (2022).

3. Awasthi, S.K., Sarsaiya, S., Kumar, V., Chaturvedi, P., Sindhu, R., Binod, P., Zhang, Z., Pandey, A., Awasthi, M.K.: Processing of municipal solid waste resources for a circular economy in China: An overview. Fuel. 317, 123478 (2022).
4. Ding, Y., Zhao, J., Liu, J.-W., Zhou, J., Cheng, L., Zhao, J., Shao, Z., Iris, Ç., Pan, B., Li, X., Hu, Z.-T.: A review of China's municipal solid waste (MSW) and comparison with international regions: Management and technologies in treatment and resource utilization. J Cleaner Prod. 293, 126144 (2021).
5. Zhou, C., Huang, N., Yang, G., Ma, S.: Assessing the sustainability of municipal solid waste management in China 1980 - 2019. Sustainable Horizons. 2, 100020 (2022).
6. Lv, J., Dong, H., Geng, Y., Li, H.: Optimization of recyclable MSW recycling network: A Chinese case of Shanghai. Waste Management. 102, 763–772 (2020).
7. Berardocco, C., Delawter, H., Putzu, T., Wolfe, L.C., Zhang, H.: Life Cycle Sustainability Assessment of Single Stream and Multi-Stream Waste Recycling Systems. Sustainability. 14, 16747 (2022).
8. Tanguay-Rioux, F., Spreutels, L., Héroux, M., Legros, R.: Mixed modeling approach for mechanical sorting processes based on physical properties of municipal solid waste. Waste Manage. (oxford). 144, 533–542 (2022).
9. Bashir, M.J.K., Chong, S.-T., Chin, Y.-T., Yusoff, M.S., Aziz, H.A.: Single Waste Stream Processing and Material Recovery Facility (MRF). In: Wang, L.K., Wang, M.-H.S., and Hung, Y.-T. (eds.) Solid Waste Engineering and Management: Volume 2. pp. 71–164. Springer International Publishing, Cham (2022).
10. Arina, D., Balodis, J., Janpaule, I., Normand, M.: Comparison of municipal solid waste characteristics after separation by star and drum screen systems. In: Civil Engineering. International Scientific Conference. Proceedings (Latvia). Latvia University of Agriculture (2013).
11. Rezaee, A., Moussavi, G., Feil, A., Norouzzadeh, R., Moradi, M., Azami, K., Keshavarz, K.: Application of disc screen for wastepaper recycling: evaluation of influential parameters. Int. J. Environ. Sci. Technol. 19, 4923–4930 (2022).
12. Lala, Z., Ang, M., Zhenfeng, Z., Siqi, J., Julong, Z.: Research Status of Cross Screen for Screening High-moisture Sticky Fine Coal. Coal Technology. 41, 244–246 (2022).
13. Huber, C., Kroisleitner, H., Stampfer, K.: Performance of a Mobile Star Screen to Improve Woodchip Quality of Forest Residues. Forests. 8, 171 (2017).
14. Bulgakov, V., Pascuzzi, S., Ivanovs, S., Ruzhylo, Z., Fedosiy, I., Santoro, F.: A New Spiral Potato Cleaner to Enhance the Removal of Impurities and Soil Clods in Potato Harvesting. Sustainability. 12, 9788 (2020).
15. Jinqing, L., Xinyu, W., Zihui, L., Jicheng, L., Zhongyuan, L.: Design and experiment of the grading device for a potato seed cutting machine. Transactions of the Chinese Society of Agricultural Engineering. 36, (2020).
16. e Silva, B.B., da Cunha, E.R., de Carvalho, R.M., Tavares, L.M.: Improvement in Roller Screening of Green Iron Ore Pellets by Statistical Analysis and Discrete Element Simulations. Miner. Process. Extr. Metall. Rev. 41, 323–334 (2020).
17. Asghari, M., Noaparast, M., Jahani Chegeni, M.: Investigation of cigarette effect and elastic-plastic behavior of green iron pellets on the roller screen efficiency. Adv. Powder Technol. 35, 104321 (2024).

The Study of Oilfield Carbon-Neutral Factory Construction

Ruihuan Zhang, Wenjin Yang, Xiangzhou Cui, Zhipeng Ba, Gaofeng Wang, Xue Sun, and Tong Wang

Abstract This study has constructed a brand-new clean and low-carbon production mode and established an oilfield as the first carbon-neutral factory for oil and gas production in China targeting Scope 1 and 2 greenhouse gas emissions. To achieve zero carbon emissions in oil and gas field enterprises through energy conservation and emission reduction measures in the production and manufacturing process is an important goal of the current transformation and development. This study systematically creates a method through energy consumption sorting, carbon verification, clean resource analysis, process energy saving optimization, scheme comparison and economic evaluation in planning layout, and implementation process of carbon-neutral factories construction. According to this method, this study sorts and optimizes the energy consumption of the central processing facility in the operation area. It utilizes solar energy, newly installs equipment such as photovoltaic module, heat pump, electromagnetic heater, and covers all the production needs with green electricity. Finally, it completely replaces the associated gas, and successfully builds carbon-neutral and achieves certification from authoritative institutions.

Keywords Low-carbon production · Renewable substitution · Carbon-neutral factory

R. Zhang (✉) · W. Yang · X. Cui · Z. Ba · G. Wang · T. Wang
Production Technology Research Institute of Yumen Oilfield Company, CNPC, Jiuquan, P. R. China

X. Sun
Refining & Chemical Plant of Yumen Oilfield Company, CNPC, Jiuquan, P. R. China

Z. Peng (ed.), *Technological Advancements and Future Directions in Green Energy*, Green Energy and Technology,
https://doi.org/10.1007/978-3-031-83203-1_6

1 Introduction

With the rapid growth of the global economy, fossil energy consumption is accelerating, and greenhouse gases [GHG] emissions are increasing, the CO2 concentration reached 415 ppm in 2020 compared to 285 ppm in 1850 [1]. Under the gradual ascending of GHG concentration in the atmosphere, the global climate is seriously affected [2, 3], the global average surface temperature rise reached about 1.2 °C in 2020 compared to 1850 [4–6]. The goal of limiting global warming to 1.5 degrees Celsius has become a global consensus, but the situation of global GHG reduction is not optimistic. At present, 138 countries already have set carbon neutrality goals, the rest have set carbon removal goals. China has set a "dual carbon" goal [7, 8], which means it will reach carbon peak by 2030 and carbon neutrality by 2060. The achievement of global carbon peak and carbon neutrality goals are related to the sustainable development of the world economy and the future of mankind [9], which is a profound economic and social change.

Under the trend of carbon goals in the world, low-carbon and zero-carbon have become an important goal of enterprise development. To control the GHG emissions, industries have begun to explore new production paths, replacing the traditional fossil energy consuming with clean energy. Currently, China's industry [10, 11], agriculture [12], and construction [13, 14] have gradually begun to adopt clean energy for transformation. At this stage, the production of the oil gas industry has begun to widely adopt clean resources such as photovoltaic, photothermal, wind power, geothermal, and other clean resources to provide energy and electricity, then to displace fossil energies gradually.

This study focuses on exploring a new production model, which has broken the traditional production mode of consuming fossil energy and constructed a carbon-neutral production mode of completely using renewable energy. According to the current classification principle, the GHG emissions in scope 3 [15] are relatively wide, and it is difficult to avoid and eradicate them. In this study, an oilfield operation area was built into China's first carbon-neutral factory [16] for oil and gas production with Scope 1 and Scope 2 GHG emissions.

2 Method and Route Study

2.1 Current Energy Consumption

This operation area is located in the Jiuquan Basin, mainly distributed in three fault blocks. Each fault block has a large difference in oil bearing properties, and there are significant differences in oil well production. The designed capacity of the central processing facility (CPF) is 190,000 tons purified oil per year, the actual capacity is currently lower than designed. The crude oil produced by each well field enters the CPF through the gathering and transportation system. The crude oil is

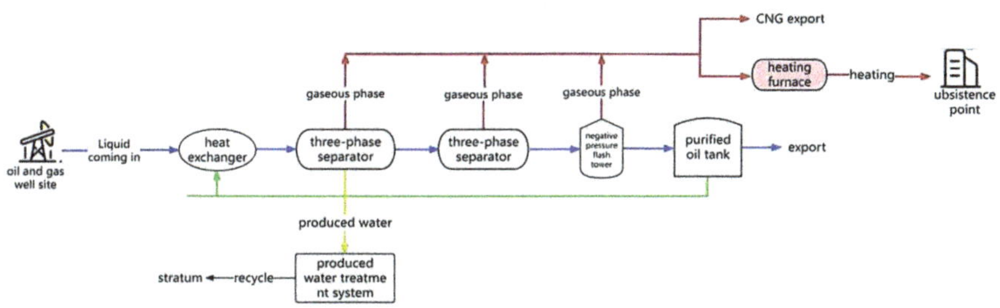

Fig. 1 Process flow diagram sketch

Table 1 Operation area energy consumption in 2022

No.	Item	Energy consuming parts	Energy consumption
1	Grid electricity	Electrical equipment, pumping units, pumps, compressors, etc.	672.042×10^4 kW·h
2	Associated gas	Oil heating, pipeline tracing, tank anti-freezing, heating, etc.	127.56×10^4 Nm^3

separated by two stages of heating and dehydration, and the purified oil is obtained. Purified oil is transported to western pipelines for sale; the extracted water is processed, then qualified water is injected back to formation; the separated associated gas goes to the associated gas treatment system and the gas boiler system for sale and burned as fuel. The process flow diagram sketch is shown in Fig. 1.

In this study, the overall energy consumption of the site is sorted out and calculated, and the two energy forms of grid electricity and associated gas are mainly used. The associated gas is used in the gas steam boiler to generate low-pressure steam. The energy consumption parts and energy consumption amount in the CPF are shown in Table 1.The overall scale of this operation area is not too large, so it has certain advantages in replacing fossil energy.

2.2 *Carbon Verification*

In this study, the physical area covered by the overall production of the operation area is taken as the accounting boundary, and it is divided into basic production system, auxiliary production system, affiliated production system. The emission factor method and material balance method are used to verify the two GHGs (CO2 and CH4) involved in the three systems.

After verification and approval by a third-party certification authority, the total GHG emission of the operation area in 2022 is 7539.21 tCO2 e. Including: emissions from fuel combustion are 2758.09 tCO2 e, emissions from industrial production processes are 297.93 tCO2 e, and emissions corresponding to net purchased electricity are 4480.57 tCO2 e. Due to the lack of effective and accurate monitoring

Table 2 Operation area GHG emission in 2022

Scope	Item	Fuel burning	Emission factor	tCO_2 e	Total emissions
Scope1	Fuel burning emissions	127.56×10^4 Nm^3	21.6219 t $CO_2e/10^4$ Nm^3	2758.09 t	7539.21 t
	Process emissions	–	–	297.93 t	
Scope2	Purchased grid electricity	672.042×10^4 kW·h	6.671 t $CO_2e/10^4$ kW·h	4483.19 t	

means for a small amount of escaped and empty associated gas, the default value method is temporarily adopted to be included in the process discharge.

It can be seen from Table 2 that in the carbon emissions of scope 1 and 2, associated gas combustion and purchased grid electricity are the main emission sources, accounting for about 96.05%. Nearly 60% of GHG emissions are indirect emissions of grid electricity. According to the data, the main GHG come from the use of grid electricity and associated gas. Therefore, the GHG emission reduction measures in this study are determined to be process energy-saving transformation, green electricity replacement, and green heat replacement.

2.3 *Renewable Energy Analysis*

After investigation, there are no available hydro-electric power resources and geothermal resources around. The wind power resource is a resource-poor area, with wind power density lower than level 1 [17] and low utilization value. Therefore, this study mainly considers the use of solar energy.

The solar energy resources of the project site were evaluated, and the sunshine hours, sunshine percentage, and total solar radiation of the weather station in the past 26 years were counted. The average annual total solar radiation was 6211.0 MJ/m2, and the average annual sunshine hours was 3347.4 h. According to the relevant provisions of the standard "Assessment method for Solar Energy resource," the annual total solar radiation in the horizontal plane of the project site is between 5040 MJ/m2 and 6300 MJ/m2, and the resource grade is B [18], which is a very resource-rich area. The total radiation stability grade of the horizontal plane is grade C [18], and the stability is general. Therefore, it has high utilization value in this zone. The specific data can be seen in Figs. 2 and 3.

This study uses solar energy as a clean alternative for the operation area. The utilization of solar energy can be reflected in photovoltaic power generation and photothermal heating. Photovoltaic power generation is a technology that uses the radiation energy of sunlight to directly convert light energy into electrical energy when the semiconductor interface produces a voltage effect. Photothermal heating is also a technology that uses the radiation energy of sunlight, relying on concentrators to focus sunlight on a specific region, resulting in an temperature increase in the

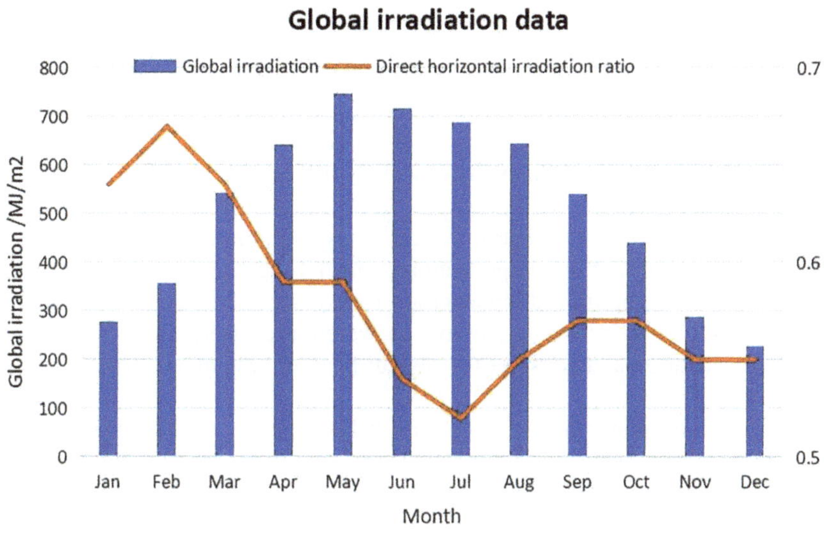

Fig. 2 Global irradiation data

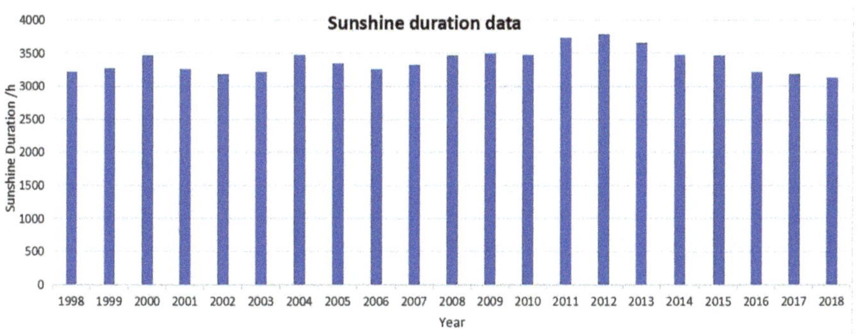

Fig. 3 Sunshine duration data

region, thereby heating the medium for heating. Both technologies [19] can meet the needs of this study and achieve clean energy supply.

2.4 Energy Optimization

In this study, the solar energy was chosen to use for the clean substitution of production status. In order to reduce the construction scale and production costs, first of all to optimize the operation area energy consumption, mainly through the following measures:

1. Pumping units: Before the optimizing, all 54 producing wells used traditional power frequency oil recovery technology, which consumed a lot of energy and had low pump efficiency. Then applied flexible frequency conversion control, intelligent pumping in 36 wells with pumping unit 45 kW rated power, after the implementation of a single pumping unit daily energy consumption can be reduced by 16~20%, and the pump efficiency increased by 2.5%. The data can be seen in Table 3.
2. Heating system: Before the optimizing, all relevant production units have set up offices here, resulting in a large heating area and high heating load. Then the heating area was optimized, concentrate part of the office, deactivate the old office buildings, concentrate part of the warehouse, use part of the garage, after optimization and consider the overall heating system heat loss data, see Table 4.
3. Steam system: Optimizing the usage period of steam in the accident tank and fire water tank, and deactivating some heat tracing pipelines, is expected to reduce steam consumption by 13.21%. The two-stage heating system is critical for production, so it is impossible to optimize. And it was separated from the original steam system. The comparison data is shown in Table 5.

After optimization, the total energy consumption of the operation area has decreased, the replacement power in the CPF has decreased from 1570.7 kW to 1404 kW, from Fig. 4 we could find the power is reduced by 10.61%. So in the subsequent replacement research, we take this as the standard of clean replacement.

Table 3 Comparison of pumping units optimization

Item	Pumping units	Electricity variation	Rate of change
Before optimization	470×10^4 kWh/a	Declined 59×10^4 kWh	Declined 16 ~ 20%
After optimization	411×10^4 kWh/a		

Table 4 Comparison of heating system optimization

Item	Heating area	Heating load	Rate of change
Before optimization	6580 m^2	814.5 kW	Declined 14.25%
After optimization	5540 m^2	698.4 kW	

Table 5 Comparison of steam system optimization

Item	Steam consumption	Equivalent power	Rate of change
Before optimization	0.53 t/h	381.6 kW	Declined 13.21%
After optimization	0.46 t/h	324 kW	

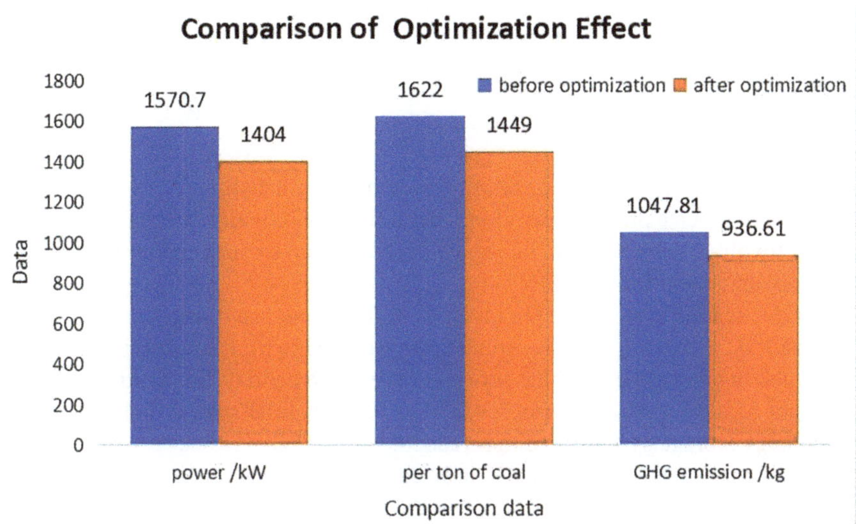

Fig. 4 Comparison of optimization effect

2.5 *Clean Alternative Technology Route*

Based on the above resource conditions, the following clean alternative technology schemes are proposed for the construction of carbon-neutral factory:

1. Second heating stage: For crude oil, this process part has a great impact on the dehydration effect and directly affects the quality of purified oil exportation. It is necessary to ensure stable production with temperature control accurate and fast. If photothermal is used, auxiliary heating equipment is still required during the non-sunlight hours. This study uses photovoltaic substitution, matching a 300 kW electromagnetic heater with temperature control. When non-sunlight shows, it can rely on grid power, which has the least impact on the current production system.
2. Steam and heating systems: They are only operated in winter, and working condition is from November to April, a total of 6 months. This part of energy can be cleanly replaced by photothermal or photovoltaic. According to the calculated power after optimization, about 1.5 MW photothermal (calculated with trough collector) or photovoltaic can meet the actual demand. From the estimated data, the investment and land area of the photothermal system have certain advantages. The specific comparison is shown in Table 6.

For the steam system of the CPF, while the new photothermal system is being built, a set of photothermal steam generation system is still needed. However, there is no effective and low-cost steam storage technology at present, so an electric steam generation system is also needed to ensure production during the non-sunlight

Table 6 Comparison of photothermal and photovoltaic construction

Item	Solar-thermal 1.5 MW	Photovoltaic 1.5 MW
Construction investment	450×10^4 CNY	780×10^4 CNY
Area covered	6200 m^2	24,000 m^2

hours. The two sets of steam generation systems significantly increase investment and also complicate process operation. Therefore, it was finally determined to adopt the photovoltaic solution and choose a 1 t/h resistance steam generator to provide low-pressure steam for the CPF.

For the heating system, the photovoltaic replacement can directly provide 85 °C hot water to the existing heating circulation system. However, considering the operation time in winter, the sunshine period is only 5~7 h per day. So in order to satisfy the night heating demand, about 2.3 MW of photothermal system and 45MWh of heat energy storage system need to be added.

This scheme can only satisfy the daily demand, but if there are continuous days of rain, snow, and other extreme weather, the photothermal system cannot continue to supply heat, and the original gas steam system still needs to be activated. Therefore, the study determines the use of photovoltaic replacement scheme and selects a 400 kW air source heat pump, which can provide 750 kW heating power. Through optimization, the total energy consumption of the operation area has decreased, and the total power of the CPF has decreased from 1570.7 kW to 1404 kW, which is about 10.61% lower than that before optimization. So we use this as a clean alternative criteria in the subsequent alternative research.

3. Photovoltaic construction scale: Ensure that the clean power supply of the newly built photovoltaic scale is greater than the original net power consumption. According to the statistics of 2022, the annual non-clean power consumption is 672.042 × 104 kW·h. First of all, based on the electricity consumption data of 2022, after considering the optimization of the reduced power consumption and the increase of the new equipment, the net increase of the annual power consumption is expected to be 385.46 × 104 kW·h. At the same time, considering the construction investment, land resources, operating costs, and other factors, the final photovoltaic construction scale is determined to be 5 MW, and the energy storage scale is determined to be 2MWh.
4. Green electricity offsetting grid electricity usage: The operation area has reached an agreement with the local electricity supply company. During daylight hours, the operation area uses the green electricity generated by its photovoltaic system while the excess green electricity is sold through the power grid. When the green electricity exported during daylight hours exceeds the non-green grid electricity used at night, the operation area achieves complete carbon neutrality in terms of the implied carbon emissions from its electricity usage.
5. Vehicles: For the GHG emissions involved in the transportation process, electric vehicles can be used to replace the original gasoline and diesel vehicles.
6. Due to the limited scale of photovoltaic construction, the production status may change during actual running, and the overall electricity consumption may

exceed the expected data, resulting in the green electricity generation not completely offsetting the non-clean power consumption. In this case, the non-clean power consumption needs to be offset by purchasing green certificates.

Also, some GHG emissions are difficult to avoid and eradicate, and these parts need to be offset by carbon trading such as China Certified Emission Reduction (CCER).

3 Discussion and Result

3.1 Optimizing Necessity

In the carbon-neutral factory construction, the original system of production should be optimized first, and the energy consumption level of the production system should be reduced. The production energy demand directly determines the scale of clean energy project and construction investment. When green power is used to replace the fossil consumption system, the power consumption will significantly be much higher than before during non-sunlight or windless periods. After the associated gas supply system was replaced with the clean energy in this area, the grid electricity used during the night and non-sunlight periods has increased a lot, leading to production costs increased. At this time, optimization can achieve good results. Compared with before optimization, the scale of new replacement projects, construction investment, power consumption, and production costs have all decreased to some extent. The specific data are shown in Table 7.

3.2 Comparison of Photovoltaic and Photothermal Method

For oil and gas production or other stations with heat demand, both photovoltaic and photothermal energy utilization methods can achieve clean replacement in the carbon-neutral factory construction, but the supporting equipment and process are different. Based on the actual operating performance of photothermal collectors and photovoltaic panels at this stage, solar-thermal conversion efficiency is higher than solar-electric conversion efficiency. It can be seen from Figs. 5 and 6 that the overall investment and land occupation of photothermal construction are better than photovoltaic construction for each megawatt.

Table 7 Value decrease after optimization

Item	Construction scale	Investment	Electricity consumption	Production costs
Decrease	0.3 MW	160×10^4CNY	77.2×10^4 kW·h	35.13×10^4CNY

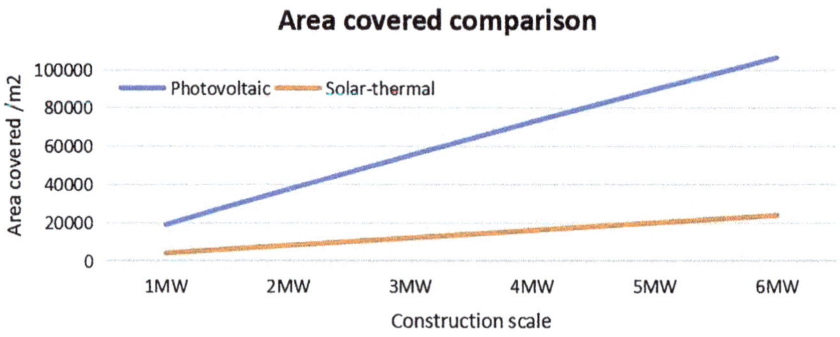

Fig. 5 Area covered comparison

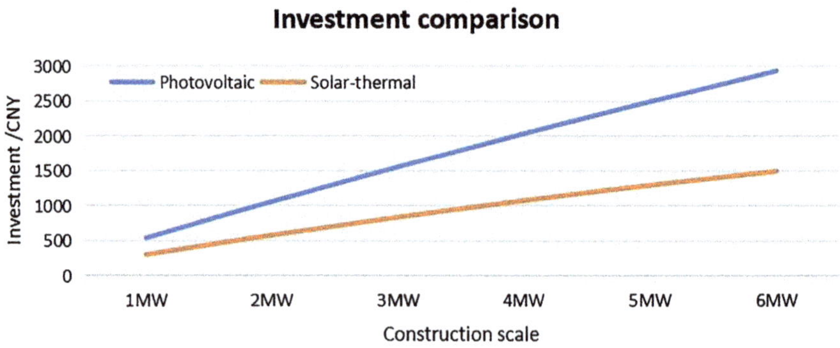

Fig. 6 Investment comparison

However, due to the instability of renewable energy sources such as solar and wind energy, it is necessary to fully consider how to ensure the normal operation of the production system during the non-clean energy time when replacing them with clean energy. If possible, a variety of new energy sources can be used complementarily to ensure the production energy supply for stable [20]. As shown in the Fig. 7, even in the months with good sunlight resources in this region, there are also poor solar resources periods in a week. At present, the price of photovoltaic modules has decreased significantly, and the power generation cost of the self-built photovoltaic power station (after considering the depreciation of the equipment) is about 0.17 CNY/kW·h, which is significantly lower than the cost of directly using the grid power 0.55 CNY/kW·h.

Since photothermal energy generation also requires emphatic considerations in ensuring that the production system can continue normal operation during the non-sunlight period. In general, photothermal replacement solutions need to be equipped with energy storage systems or another set of production facilities that rely on grid

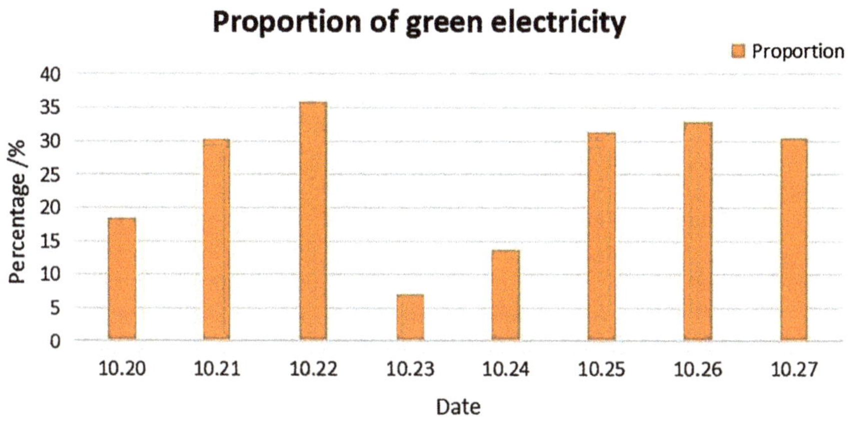

Fig. 7 Proportion of green electricity

Table 8 Construction scheme comparison

Item	Satisfy sunlight period demand	Satisfy full day demand
Construction scale	1.5 MW	6.0 MW
Energy storage scale	0	30MWh
Construction investment	450×10^4 CNY	1800×10^4 CNY
Area covered	6200 m^2	25,000 m^2

power. When matching energy storage systems, it is necessary to expand the scale of photothermal construction to ensure that the stored energy during the daytime can meet the demand during the non-sunlight period. At this time, the investment will increase to a large extent. Taking into account the actual situation of 1.5 MW in this operation area, the average daily sunshine time in winter is only 6 hours. If relying on photothermal energy to meet the daily production needs, it is still necessary to build a another 4.5 MW photothermal construction and 30MWh heat storage construction scale. As shown in Table 8, the total investment and land use are significantly increased, but if there are consecutive extreme weather conditions, production still cannot be guaranteed.

When accompanied with another production system, it will make the production process more complex, and a highly efficient and rapid control system needs to be designed to switch between the two production systems. When the original system is not completely deactivated, and only partial replacement is considered, clean photothermal alternatives have certain advantages. Therefore, whether to adopt photothermal replacement or photovoltaic replacement, it needs to be comprehensively considered according to the actual production energy consumption amount, consumption period, and energy form.

3.3 Equipment Selection

Ideally, when building a carbon-neutral factory by replacing fossil energy with electricity, we should try to control the overall cost of electricity to be similar to the cost of fossil energy consumption, so as to ensure that the original production cost will not be greatly affected. In most regions, natural gas costs are lower than grid electricity costs under the same heat supply conditions, so the production cost of electromagnetic or resistance heating equipment is higher than that of associated gas heating equipment. We can find in Fig. 8 that when the local natural gas price 2.8CNY/Nm^3 and the average calorific value 35.4 MJ/Nm^3 are taken into calculation, providing the same amount of heat, converting gas price to the electricity costs, the electricity price is only one-tenth of the gas price, converting the electricity price to the gas costs, the natural gas price will be twice as high as the local gas price. This means that the production cost of using grid electricity is about twice higher than using gas. At this time, suitable equipment and appropriate processes for this equipment will have a greater impact on production costs.

For traditional equipment that consumes electricity to generate heat, even if the efficiency of electric-thermal conversion of the equipment is high, it is impossible to fully convert 1 kW·h electricity to 3.6×10^6 J heat. When using an air source CO_2 heat pump with the coefficiency of performance (COP) > 2.1 at −25 °C, even considering heat loss, 1 kW·h electricity can be converted to more than 7.2×10^6 J heat. Therefore, after using a heat pump, power consumption can be reduced by more than 50%. Higher the temperature is, more heat the heat pump can output for each kW·h consumed, which can be seen from Fig. 9. However, due to the constraints of the circulating medium, the heat pump can only provide heat at 70–80 °C, which has certain limitations in application and is suitable for heating and low-temperature demand systems.

For the summer operating conditions in this operation area, the electromagnetic heater can be suspended, and the heat pump can be used to provide second stage heat for the temperature rising of crude oil. According to the characteristics of the heat

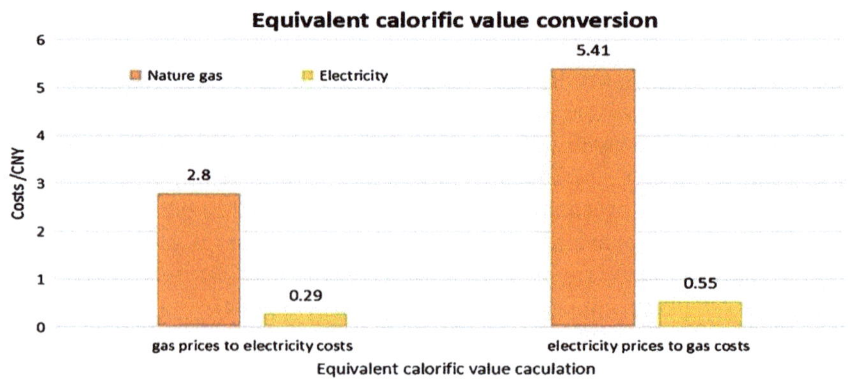

Fig. 8 Equivalent calorific value conversion

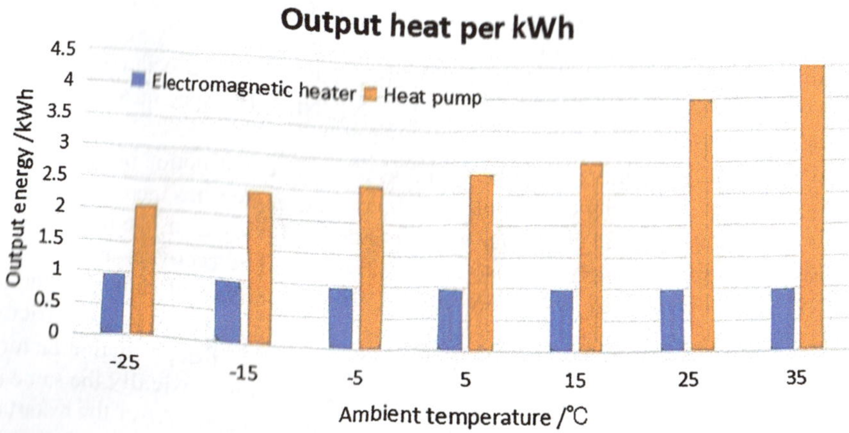

Fig. 9 Output heat per kWh

Table 9 Equipment scheme comparison

Item	Electromagnetic heater	Heat pump	Rate of change
Electricity consumption	117.6 × 10^4 kW·h	42 × 10^4 kW·h	Declined 66.67%
Cost comparison	53.51 × 10^4CNY	19.11 × 10^4CNY	

pump, when the temperature is above 15 °C, the COP can be greater than 3, and when the temperature is above 25 °C, the COP can be greater than 4. Taking into account the temperature changes, the average COP = 3 is used for energy estimation. The annual photovoltaic green electricity consumption is calculated at 25%, and the average price of 0.17 CNY/kW·h is considered. The annual use of non-green grid electricity is calculated at 75%, and the average price of 0.55 CNY/kW·h is considered. At this time, production energy consumption can be further reduced. The specific changes in production costs when only considering electricity prices are shown in Table 9.

3.4 Benefit Analysis

Due to the fact that this study only adjusted some processes and energy access methods, the original production conditions did not change, and management and operating personnel did not change either, this study temporarily ignored other production cost factors and only considered and demonstrated the changes in production costs for the energy-consuming part of the operation area after clean substitution.

The annual generation of 5 MW photovoltaic is about 956.5 × 10^4 kW·h, and the annual estimated total consumption is 991.48 × 10^4 kW·h, including 248.14 × 10^4 kW·h green electricity and 744.42 × 10^4 kW·h grid electricity. The average price of green electricity is 0.17 CNY/kW·h, and the average price of grid electricity is 0.55 CNY/kW·h. The green electricity sales on the grid is

708.36 × 10^4 kW·h, with an average price of 0.3078 CNY/kW·h. The reduced consumption of associated gas in the operation area can be completely separated into natural gas and mixed hydrocarbons for external sales by the associated gas treatment unit, with an average selling price of 2.3 CNY/Nm^3. The specific analysis data can be seen in Table 10.

After overall estimation, it can be seen that after the construction of clean alternatives, the total electricity consumption in the operation area has increased significantly, and the use of grid electricity has also increased. However, the use of clean electricity and power supply has also enhanced greatly. The recovery and external sales of all associated gas have been attained, and the increased sales revenue of green electricity and associated gas is higher than the cost of using grid electricity. After considering the cost of newly purchased equipment and depreciation factors, the production cost when realizing clean alternatives remains basically the same as before, and the economic benefits have not been greatly affected. After the restart of domestic CCER trading in the future, the operation area can participate in CCER trading with reduced carbon emissions to increase income.

4 Conclusion

This study ultimately achieves the generalized clean and low carbon substitution of the operation area, after optimizing the production process and building 5 MW photovoltaic power plant, and adopting air source heat pump, electromagnetic heater, resistance steam generator. When the carbon-neutral factory was completed, it was certified by an authoritative third-party organization and issued a standard-level carbon-neutral factory certification. At the same time, it can provide green electricity 956.5 × 10^4 kW·h per year, recover associated gas about 127.56 × 10^4 Nm^3, reduce CO_2 emissions 7539.21 t, with good demonstration effect and social benefits.

Through this study, the following basic route for the establishment of carbon-neutral factories is also obtained: (1) Fully identify the actual production status of the factory, conduct a comprehensive and detailed carbon verification, then identify the surrounding available and valuable clean energy sources, finally develop an alternative plan that matches the actual production. (2) When using clean energy to substitute, the original system should be optimized for energy conservation first, consider the characteristics of production and resources, combine production and resources efficiently, compare multiple plans and find out the most suitable alternative plan. (3) Consider the most suitable equipment, choose more efficient and energy-saving equipment as much as possible on the premise of meeting the production requirements. (4) Before construction, a detailed economic benefit analysis should be conducted to achieve clean alternative and low-carbon production with the least impact on the factory's benefits and the least increase in production costs. Therefore, we believe that the new production mode studied in this paper has certain reference value for many factories' future clean transformation and contributes to the realization of the global dual carbon goals.

Table 10 Benefit analysis

Item	Original condition	Substituted condition	Value change	Costs change
Total electricity	672.04×10^4 kW·h	992.56×10^4 kW·h	$+320.52 \times 10^4$ kW·h	/
Green electricity consumption	0	248.14×10^4 kW·h	$+248.14 \times 10^4$ kW·h	$+42.18 \times 10^4$CNY
Grid electricity consumption	672.04×10^4 kW·h	744.42×10^4 kW·h	-262.61×10^4 kW·h	$+409.43 \times 10^4$CNY
Green electricity sale	0	708.36×10^4 kW·h	$+708.36 \times 10^4$ kW·h	-218.03×10^4CNY
Associated gas consumption	127.56×10^4 Nm^3/a	0	-127.56×10^4 Nm^3/a	-292.86×10^4CNY
Production costs	369.62×10^4CNY	310.34×10^4CNY	/	-59.28×10^4CNY

Note: Compared to the original economic analysis, data has been simplified, but the final conclusion remains the same

References

1. Chen, J.M. Carbon neutrality: Toward a sustainable future. Innovation 2021, 2, 100127.
2. Solomon S, Qin D, Manning M, et al. Climate Change 2007: The Physical Science Basis (IPCC) [Z]. Cambridge University Press. 2007.
3. Field C B. Managing the risks of extreme events and disasters to advance climate change adaptation: special report of the intergovernmental panel on climate change [M]. Cambridge University Press, 2012.
4. Matthews, H.D.; Wynes, S. Current global efforts are insufficient to limit warming to 1.5 °C. Science 2022, 376, 1404–1409.
5. Meinshausen, M.; Lewis, J.; McGlade, C.; Gütschow, J.; Nicholls, Z.; Burdon, R.; Cozzi, L.; Hackmann, B. Realization of Paris Agreement pledges may limit warming just below 2 °C. Nature 2022, 604, 304–309.
6. Yan, M.; Liu, J.; Wang, Z.; Ning, L. Biogeophysical impacts of land use/land cover change on 20th century anthropogenic climate compared to the impacts of greenhouse gas change. Int. J. Climatol. 2020, 40, 6560–6573.
7. The State Council of China. Opinions on Fully Implementing the New Development Philosophy and Doing well in Carbon Emissions Peaking and Carbon Neutrality Work. 2021. Available online: https://www.gov.cn/zhengce/2021-10/24/content_5644613.htm.
8. The State Council of China. Notice on Issuing the Action Plan for Carbon Emissions Peaking Before 2030. https://www.gov.cn/gongbao/content/2021/content_5649731.htm
9. Arsad A Z, Hannan M, Al-Shetwi Ali Q, et al. Hydrogen energy storage integrated hybrid renewable energy systems: a review analysis for future research directions. Int J Hydrogen Energy 2022. https://doi.org/10.1016/j.ijhydene.2022.03.208.
10. Chen C, Wang M, Chen X, et al. Recent progress in solar photothermal steam technology for water purification and energy utilization, Chemical Engineering Journal, Volume 448, 2022, 137603, ISSN 1385–8947, https://doi.org/10.1016/j.cej.2022.137603.
11. Li F, Yu Y, Shu Y, et al. Study on characteristics of photovoltaic and photothermal coupling compressed air energy storage system, Process Safety and Environmental Protection, Volume 178, 2023, Pages 147–155, ISSN 0957-5820, https://doi.org/10.1016/j.psep.2023.08.033.
12. Wu, G.; Fang, H.; Zhang, Y.; Li, K.; Xu, D. Photothermal and Photovoltaic Utilization for Improving the Thermal Environment of Chinese Solar Greenhouses: A Review. Energies 2023, 16, 6816. https://doi.org/10.3390/en16196816
13. Cui Y, Zhang X. Research and analysis of energy consumption and energy saving in buildings based on photovoltaic photothermal integration. Scientific Reports. 2024 Jan 9;14(1):923. doi: https://doi.org/10.1038/s41598-024-51209-1. PMID: 38195975; PMCID: PMC10776761.
14. Li Y, Yang W, Ling Z, et al. Research progress in solar energy utilization technology in buildings. Refrig. Air Condition. 2022;22(08):92–97.
15. ISO 14064-1: Greenhouse gases—Part 1: Specification with guidance at the organization level for quantification and reporting of greenhouse gas emissions and removals[s]. Switzerland, ISO copyright office, 2018.
16. T/CAS 764-2023: Zero-carbon factory creation and block chain-based evaluation specifications[s]. Beijing, CAS, 2023.
17. NB/T 31147-2018, Technical Code for Wind Energy Resource Measurement and Assessment of Wind Power Projects[s], Beijing, China Water & Power Press, 2018.
18. GB/T 37526-2019, Assessment method for solar energy resource[s], Beijing, Standards Press of China, 2019.
19. Li Y, Zhang T, Zhang H, et al. Efficient and comprehensive utilization technology of solar photovoltaic photothermal energy. Power Gen. Technol. 2022;43(03):373–391.
20. J. Jurasz, F.A. Canales, A. Kies, M. Guezgouz, A. Beluco, A review on the complementarity of renewable energy sources: Concept, metrics, application and future research directions, Solar Energy, Volume 195, 2020, Pages 703–724, ISSN 0038-092X, https://doi.org/10.1016/j.solener.2019.11.087.

Environmental Analysis of Oxygen Liquefaction Integrated with Green Hydrogen Production for Enhanced Economic Feasibility

Wagd Ajeeb, Ricardo Assunção, and Rui Costa Neto

Abstract Promoting the use of renewable energies and reducing the emissions caused by the consumption of fossil fuels have become significant points for the global industrial community. Green hydrogen, produced from renewable sources, is a promising energy vector, but its implementation has been hurdled by its cost premium in relation to grey hydrogen. One way to increase the economic feasibility of green hydrogen is to market the byproduct O_2. In our previous work, an oxygen liquefaction process was developed to be used in a water electrolysis plant that can output liquid oxygen (LO_2) with lower specific energy consumption and cost than other O_2 production methods like cryogenic distillation (CD) and pressure swing adsorption (PSA). In this study, life cycle analysis is conducted for a 50 MW water electrolysis plant with an integrated oxygen liquefaction unit aiming at assessing its global warming potential (GWP) in four different energy supply scenarios of the Portuguese context. The energy and material balance were determined by simulation with the Aspen Plus software and then GWP impact was assessed using SimaPro software. The results demonstrated that adding an oxygen liquefaction unit to the electrolysis plant only increased the GWP by 6.5–6.8%, compared to the isolated water electrolysis plant. Additionally, the GWP of the O_2 liquefied in the unit proved to be inferior by 34–81%, depending on the energy source, when compared to the LO_2 produced at air separation units. Lastly, the cost analysis showed that for the Portuguese case, the 50/50 combination of solar photovoltaic (PV) and onshore wind energy outputs similar levelized cost of H_2 (LCOH) and levelized cost of LO_2 (LCOLO2) as the current grid mix, being the lower costs achieved when the contribution of solar is higher, reaching LCOH equal to 3.37 €/kg_{H2} and LCOLO2 equal to 0.058€/kg_{O2} for 80PV/20Wind ratio.

Keywords Green H_2 · Oxygen liquefaction · GWP · LCA · CO_2 emissions

W. Ajeeb (✉) · R. Assunção · R. C. Neto
IN+ Center for Innovation, Technology and Policy Research, IST-ID -Associação do Instituto Superior Técnico para a Investigação e Desenvolvimento, University of Lisbon, Lisbon, Portugal
e-mail: wagdajeeb@tecnico.ulisboa.pt

Z. Peng (ed.), *Technological Advancements and Future Directions in Green Energy*, Green Energy and Technology,
https://doi.org/10.1007/978-3-031-83203-1_7

1 Introduction

Renewable energy sources (RES), such as wind and solar power, are becoming more widespread, continuously improving in efficiency, storage capacity, and in economic viability [1, 2]. Green hydrogen is a promising energy carrier to bridge the gap between RES and hard to electrify sectors, that can be produced sustainably using water electrolysis [3]. Among others, green hydrogen can be used for many applications such as chemical [4] and industrial production, transportation, power generation, grid balancing, heat generation, etc. [5, 6]. Even though hydrogen generation by water electrolysis is a very energy intensive process, in recent years there was a noticeable reduction in the levelized cost of energy (LCOE) of RES, making them viable to fuel such process and further contribute to the replacement of fossil fuels [1, 7]. The progress in hydrogen production methods aligned with advancements in renewable energy technology offers a promising landscape for the promotion of sustainable generation of green hydrogen and a cleaner environment less reliant on fossil fuels [8, 9].

During hydrogen generation by electrolysis, molecular oxygen is also formed and often discarded. However, various industries such as medical, pulp and paper, and others that use high-temperature furnaces with oxyfuel combustion utilize oxygen as feedstock [10]. This oxygen is generally bought from air separation units (ASU), which perform the cryogenic distillation (CD) of air, or produce it onsite using pressure swing adsorption (PSA) [11]. The oxygen produced during electrolysis is of similar or even higher purity than the O_2 produced via PSA and CD and corresponds to half the volume and eight times the weight of the hydrogen produced. Given that the price of liquid oxygen ranged from 0.091 €/kg_{O2} and 0.193 €/kg_{O2} in 2022 for most industrial applications, and 0.68 €/kg_{O2} for bulk medical oxygen, the losses in revenue could be great if the electrolytic oxygen would otherwise be released into the atmosphere [12, 13].

Hydrogen production plants without nearby oxygen consumers have this economic potential but also the challenge to overcome the spatial barrier between the production site and the consumer. To deal with this hurdle, in our previous work an O_2 liquefaction process was developed, simulated in the software Aspen Plus v11, that can be implemented in green hydrogen production plants to liquefy the byproduct O_2, so that it can be more efficiently stored and transported to the consumer [14].

Given the current panorama and the future roadmap, it is now an advantageous moment to utilize renewable energy sources for green hydrogen production, and coupling the H_2 production with an O_2 liquefaction unit may bring more economic feasibility to the projects by reducing the cost of green H_2 and possibly increasing the environmental advantages which this study addresses here [15, 16]. Therefore, a life cycle analysis of a 50 MW electrolysis plant with the incorporation of an O_2 liquefaction unit as described in [14] was performed. An accurate technical assessment was done before conducting the LCA studies on systems associated with hydrogen production, especially to collect the needed data related to system components and the consumption of materials and energy during the project's lifetime. A precise analysis of the oxygen liquefaction unit connected to hydrogen electrolysis powered by RES source (wind and PV solar) may improve the economic viability as well as lowering the overall environmental performance (e.g., global warming potential [GWP] impact) [17].

2 Methods and Materials

In this study, an environmental and economic analysis is carried out on the integration of oxygen liquefaction systems into a green hydrogen production plant. Firstly, the technical aspects of the oxygen liquefaction system from [14] are evaluated using Aspen Plus v11 software, defining the energy amounts and material requirements for the unit construction and process operation. Furthermore, the information regarding hydrogen production through electrolyzer was compiled from literature, project sites, and company producers. Then, life cycle analysis (LCA) is conducted to evaluate the global warming potential (GWP) impact of the total process using the SimaPro software, including hydrogen production and oxygen liquefaction. For this, different energy sources were used, such as the Portuguese grid mix and different ratios of solar photovoltaic and wind energy. The GWP environmental results were analyzed, and the respective economic analysis was performed for each of the different energy sources, for both, the production of hydrogen and the liquefaction of oxygen. Therefore, this study analyzes the GWP impact of hydrogen production by electrolysis in a 50 MW plant along with the GWP associated with the respectively sized oxygen liquefaction operation, being both variable dependent on the various energy sources considered.

2.1 The System: Hydrogen Production and Oxygen Liquefaction

In previous work by the research team, an oxygen liquefaction process was presented, designed to liquefy a stream of 100% O_2 originating from electrolysis as a byproduct during H_2 production [14]. The process presented schematically in Fig. 1, was developed in the Aspen Plus v11 software, and was presented in different scales to fit electrolysis plants from 5 to 500 MW. Here, it is intended to further analyze the 50 MW electrolysis plant and the liquefaction unit sized to fit a 50 MW electrolysis plant.

The electrolysis plant of 50 MW was considered to use conventional Alkaline electrolyzers of 60% efficiency (lower heating value) and have an energy consumption of 55 kWh per kg H_2, with a 10 L of deionized water consumption per kg H_2. The annual operating hours of the electrolyzer system were estimated to be 6600 h/year, producing a total of 6000 t_{H2}/y and 48,000 t_{O2}/y. The oxygen liquefaction unit

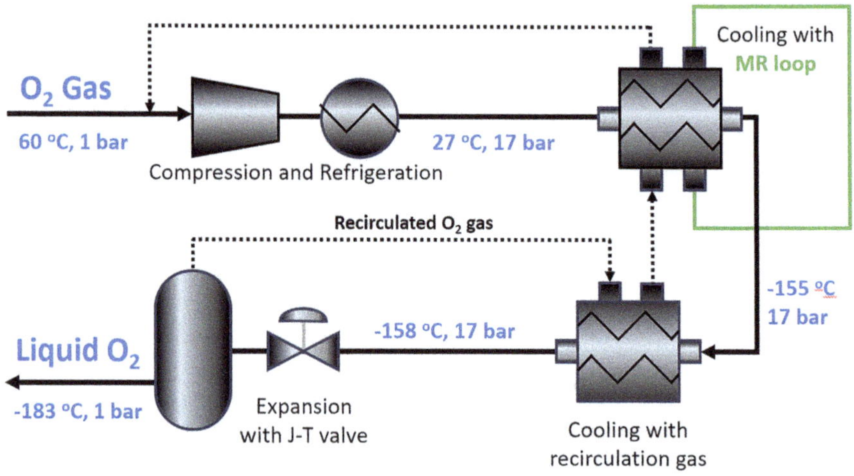

Fig. 1 Schematic representation of the O_2 liquefaction process. (Adapted from [14])

(Fig. 1) would work in continuous operation, running 8760 h/y and liquefying oxygen with a specific energy consumption of 445 kWh/tO_2. The oxygen produced from the electrolysis system is assumed to be directly fed to the oxygen liquefaction system (with brief water/H_2 removal and intermediate storage in appropriate tanks).

2.2 *LCA Cycle Inventory*

The LCA study is carried out following the ISO 14044 requirements [18] and relevant guidelines for hydrogen production and associated systems [19]. The LCA study involves the four accredited stages: goal and scope determination, inventory assessment, impact analysis, and interpretation of the results. The Life Cycle Assessment (LCA) study utilizes the ILCD 2011 Midpoint assessment methodology, aligning with recommendations from the European Commission [20]. The assessment focuses on the production of 1 kg of liquid oxygen as a functional unit, as well as 1 kg of hydrogen.

Hydrogen production The operation boundaries of the water electrolysis system are important, encompassing both the inputs (materials and processes) and process stages of the construction stage, operation stage and waste treatment stage. The construction stage involves the two main units: the energy source system (hydropower plants, PV solar panels, wind turbines or the grid electricity mix) and water electrolysis system. Each unit encompasses raw materials acquisition, manufacturing, transportation, operation and maintenance, dismantling, and waste treatment. The used electrolyser in the current study is a ALE type which typically consists of two nickel electrodes separated by a membrane layer and an alkaline electrolyte liquid [21]. In the cathode side, hydroxide ions (OH−) are generated and migrate through the membrane to the anode side, where hydrogen is produced [22]. Alkaline electrolysis operates at relatively low temperatures, typically ranging from 30 to 80 oC, and utilizes readily available and cost-effective materials for its electrolyte solution, membrane, and electrodes[23, 24]. According to the information collected for ALE, common materials such as nickel, steel, aluminium, copper and rubber are used for the electrolyser components in the cell frame, gasket and cathodes, as well as the Zirfon membranes. Also, alongside the stack components, the Balance of Plant (BoP) is taken into account in the current LCA. This includes tanks, heat exchangers, pumps, power electronics/inverters, potassium hydroxide filters, and gas separators heat exchangers such as pool boiling [25], [26] or compact heat exchangers [27], [28] and advanced thermal fluid [29].

Through inventory analysis using SimaPro software (version 9.3.0.3), the defined inputs and outputs from the electrolysis process were assessed and categorized based on GWP impact. The operation inputs data for both the RES plant and electrolyzer system including water, potassium hydroxide (KOH), cleaning materials, maintenance requirements, weather, and industrial conditions (e.g., solar irradiation and wind speed), are mainly sourced from the Portuguese database. In cases where Portuguese data is unavailable, European databases are considered for evaluation.

Oxygen liquefaction As previously mentioned in the context of LCA for hydrogen production, the operational boundaries for the oxygen liquefaction system are also defined. These boundaries encompass both the inputs, which include materials and processes, and the various stages of the process. The construction phase of the oxygen liquefaction involves raw materials, manufacturing, transportation, maintenance, and waste treatment. The oxygen liquefaction unit was previously simulated in the software Aspen Plus v11 [14]. From the simulation, it is possible to determine the weight of each equipment composing the liquefaction unit as well as their specific installed weight (component weight + structure and ancillary systems). From there, a total weight of 195 t was determined for the overall system. In Fig. 2 it is possible to see the distribution of the materials that go into the construction of the O_2 liquefaction unit. The aluminium alloy is mostly used for the two heat exchangers and the stainless steel for most of other systems like compressors, piping, valves, coolers, and the flash separator. For structure, mostly carbon steel is used with a smaller fraction of concrete. PVC is representative of most plastic parts and polyurethane is used for thermal insulation. Copper is used for electrical systems and refractory ceramics are used in places where the temperature can be higher which in this case of this process would be in the area below the compressors. The mixed refrigerant works in a closed loop and is composed of a mixture of small hydrocarbons and N_2, with a specific composition optimized to fit the cooling needs [14].

Electricity Supply As it is preferred for green hydrogen projects to use solar PV and wind energy to power their production, this study considered that powering the electrolysis system mainly would be made by a ratio of solar PV and wind energy (onshore). The three ratios studied were the 20/80, 50/50 and 80/20 PV/w power

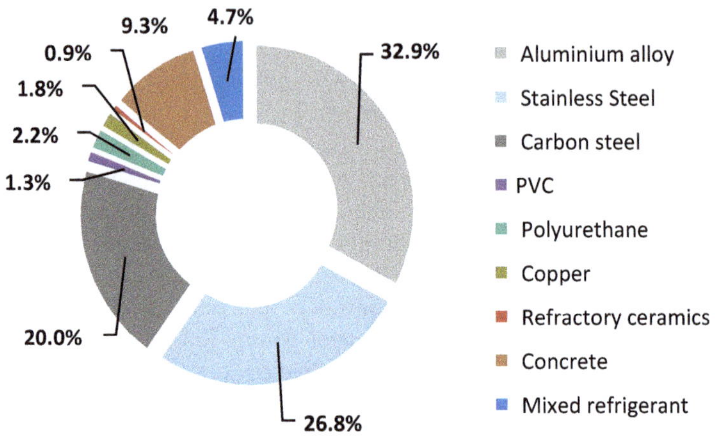

Fig. 2 Distribution of the materials of the components that make up the total 195 tonnes of installed weight of the O_2 liquefaction unit

scenarios, for comparison, the Portuguese electrical grid energy mix was also used as a fourth scenario. Each scenario considers the total capacity of 50 MW of electrolysis process distributed by the various Alkaline stacks. Electricity production data are defined from the Ecoinvent version 3 database in SimaPro, reflecting the conditions and performance of each RES in Portugal. PV solar panels are selected in the SimaPro database for low-voltage electricity generation, with a 570 kWp utilizing Si PV panels and an inverter. The onshore wind turbines chosen for this study have a capacity of 4.5 MW Enercon E-112 type, which includes infrastructure elements such as towers, blades and electricity grid [30]. On the other hand, the electricity grid mix in Portugal is considered for the data collected in 2023 and modelled in SimaPro software, where it involves 92% of electricity generated in Portugal from different resources (27% Wind, 10% Solar, 30% Hydropower, 8% Bioenergy, 21% Natural gas and 4% other fossil fuels) [31] and 8% of electricity is imported from Spain [32].

In SimaPro, the inventory of wind turbines and PV panels were modelled for the location of Portugal, considering the production of high voltage 1 kWh of electricity distributed to the grid, including operation and maintenance activities with the periodic replacement of materials.

Waste treatment The waste treatment phase for the components after a 20-year lifetime followed the strategy outlined by Lotric et al. [33] for both the oxygen liquefaction and hydrogen production systems. Recycling rates for each material (steel, nickel, plastics, copper) were estimated based on information from the recycling industry and available literature. For instance, steel was estimated to have a 60% reuse rate, while aluminium had a 96% recycling rate. Additionally, most other materials within the system were assumed to have a 99% recycling rate, with a 1%

rate allocated for landfill disposal. The option of landfill disposal was only considered when material recovery or recycling was deemed impractical.

2.3 *Economic Analysis Procedure*

To understand the economic viability of a hydrogen production plant of 50 MW with an integrated O_2 liquefaction unit the levelized cost of hydrogen (LCOH) was calculated based on the total costs related to the water electrolysis operation. On the other hand, the levelized cost of liquid oxygen (LCOLO2) was calculated assuming that the production of oxygen by electrolysis comes at no cost, as it is a byproduct, which means that only the liquefaction costs were considered for its calculation.

Both parameters were calculated based on the net present value estimation at a discount rate of 6% as previously described in [14, 34], although here the project was considered to have a length of 20 years, where an investment of an extra 30% of the total CAPEX was calculated for the electrolyzers' stack replacement at the beginning of the 11th year. The CAPEX for the electrolyzers was set as 607,500 €/MW with an extra 30% of the total CAPEX allocated to EPC and the OPEX (excluding electricity) was set as 5% of the total CAPEX. Likewise, an extra 20% of the total CAPEX of the liquefaction unit was also considered to be invested at the beginning of the 11th year for refurbishing maintenance, which would replace various parts (in compressors and other systems), to ensure an efficient operation for the second 10 years. The total CAPEX for the O_2 liquefaction unit was 8.83 M€ with an annual OPEX (excluding electricity) of 6%, and the unit power was 2.44 MW.

The electricity costs for the Portuguese grid mix were determined based on the market results for the past 6 months (Nov 23–April 24) which averaged at 47 €/MWh. The solar PV and onshore wind electricity costs in the Portuguese context were taken from [35] and were respectively 43 and 51 €/MWh.

3 Results and Discussion

3.1 *GWP Impact of Total LCA*

GWP impact has been considered a significant impact in LCA studies representing the GWG emissions. For the current study, the validation of the modelled RES types (the four scenarios) was validated primarily according to the GWP impact reported in related literature, such as between 8 to 62 g CO_2-eq./kWh for wind turbines and 132 g CO_2-eq./kWh for PV systems [36], 9.7 g CO_2-eq./kWh for onshore wind turbines [37], and between 12 to 17.6 g CO_2-eq./kWh for solar PV in the EU [38]. Figure 3 presents the results of the total GWP impact from the whole operation,

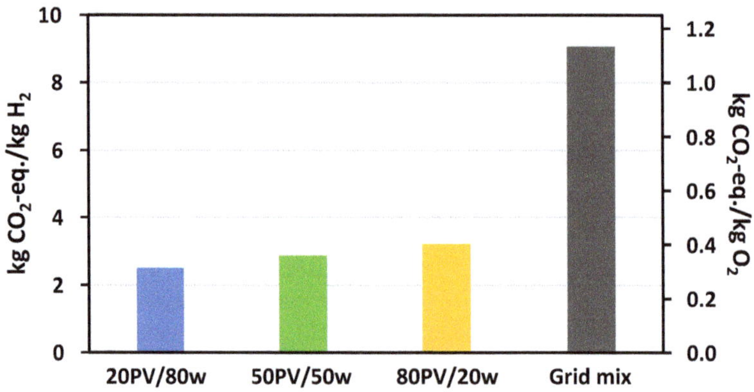

Fig. 3 GWP in kgCO_2-eq per kg O_2 and per kg of H_2 for the total LCA of the whole operation for the four different electricity supply scenarios

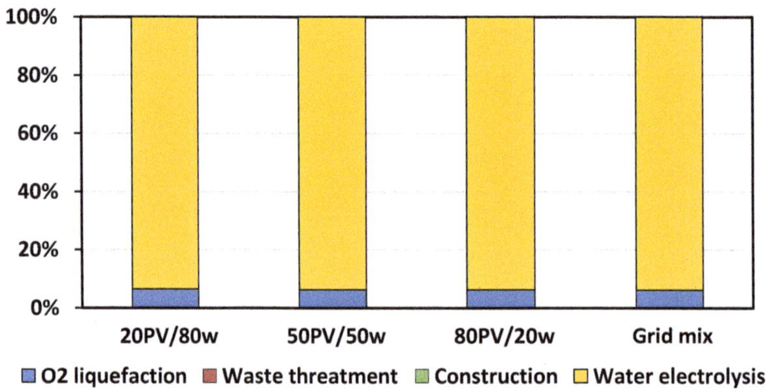

Fig. 4 GWP in percentages for the total LCA of the total operation (Water electrolysis + O_2 liquefaction) for each of the four different scenarios

water electrolysis followed by oxygen liquefaction, including all the stages for the four electricity supply scenarios.

The total GWP of the total LCA (considering all stages) was the lowest when using 20PV/80w electricity, followed by the 50/50 scenario, and then by the 80PV/20w scenario. For the grid mix scenario, the GWP was the highest, being between 2.8 to 3.6 times higher than the three renewable scenarios. Furthermore, the contribution of each of the different stages in the life cycle (construction, operation, and waste treatment) was shown in Fig. 4 as a percentage of the total GWP determined. The results show that the different electricity supplies of the four

Table 1 GWP for the total LCA (per kg H_2 and kg O_2) and for the water electrolysis stage and the O_2 liquefaction stage

	Water electrolysis stage	O_2 liquefaction Stage	Total LCA electrolysis and O_2 liquefaction	
Units	kg CO_2-eq./kg H_2	kg CO_2-eq./kg O_2	kg CO_2-eq./kg O_2	kg CO_2-eq./kg H_2
20PV/80w	2.35	0.020	0.31	2.51
50PV/50w	2.69	0.023	0.36	2.87
80PV/20w	3.03	0.025	0.40	3.23
Grid mix	8.51	0.069	1.13	9.06

scenarios didn't significantly change the GWP contributions by each stage. The major GWP contribution (~94%) comes from the water electrolysis stage, followed by the O_2 liquefaction phase (~6%). Both these processes are energy intensive, being the electricity consumed by the main contributor for these two stages. For the construction and waste treatment stages, their combined GWP is below 1%.

The absolute values obtained for the GWP for each of the four scenarios can be seen in Table 1, as a total LCA and for the independent stages of electrolysis and O_2 liquefaction. The results of GWP for the water electrolysis stage determined are not far from related literature, such as 0.68 kg CO_2-eq./kg H_2 using wind energy [39] and 3.08 kg CO_2-eq./kg H_2 using PV solar [40]. There, it is also possible to see that the integration of a liquefaction unit in the electrolysis plant only increases the overall GWP by 6.5–6.8%, depending on the electricity source. The GWP of bulk liquid oxygen from an ASU was previously reported to be 0.105 kg CO_2-eq./kg O_2 [41] (subtracting the transport contribution which would add 0.022 kg CO_2-eq./kg O_2 for a 260 km distance). Comparing with the results obtained for the GWP in this work for the O_2 liquefaction, it can be seen that for the 20PV/80w scenario, the GWP was 81% lower, 78% lower for the 50PV/50w scenario, and 76% lower for the 80PV/20w scenario. For the grid mix scenario, the GWP was still 34% lower than the reference value for the bulk liquid oxygen produced at an ASU.

3.2 *Economic Analysis of the H_2 Production with Integrated O_2 Liquefaction*

The levelized costs of H_2 and of liquid O_2 were calculated for each of the four scenarios and the results are shown in Table 2. The outcome shows that almost equal results are obtained for the 50PV/50w and the grid mix. As the LCOE of wind taken from [35] was higher that the solar, the results point out that for Portuguese projects, further integration of solar in the energy ratio can lower slightly the LCOH and LCOLO2.

Table 2 Levelized costs of hydrogen and liquid O_2 based on the LCOE of each scenario

	LCOE (€/MWh)	LCOH (€/kg_{H2})	LCOLO2 (€/kg_{LO2})
20PV/80w	49.4	3.63	0.060
50PV/50w	47.0	3.50	0.059
80PV/20w	44.6	3.37	0.058
Grid mix	47.0	3.50	0.059

4 Conclusion

In this work, life cycle analysis was performed to determine the GWP associated with our previously proposed process of O_2 liquefaction integrated in a water electrolysis plant. It was verified, in this study, that powering the total operation with a mix of solar and wind could reduce in 64–72% the GWP when compared with the Portuguese grid mix. Furthermore, it was determined that the integration of an O_2 liquefaction unit would only increase the overall GWP by 6.5–6.8%, when compared to the standalone electrolysis. Moreover, the liquefaction process could reduce the GWP associated with the production of LO_2 when compared with standard production of LO_2 methods like cryogenic distillation at air separation units, on top of the lower cost and lower specific energy consumption already reported before [14]. The reduction in GWP was dependent on the energy source, being in the worst case less by 34% (Portuguese grid mix) and in the best case analyzed less by 81% (20PV/80w). The short economic analysis showed that for Portugal, an electricity supply of combined solar PV and onshore wind can achieve lower costs than the grid mix (LCOH = 3.50 €/kg_{H2}) for the cases where solar has a higher ratio than wind, e.g., LCOH = 3.37€/kg_{H2} at 80PV/20wind. On the contrary of economic analysis, the GWP value was increasing by increasing the solar percentage in PV/Wind ratio up to 3.23 kg CO_2-eq./kg H2 but still much lower than the electricity mix scenario of 9.06 kg CO_2-eq./kg H_2.

References

1. "Our World in Data. Renewable Energy," https://ourworldindata.org/renewable-energy, 2023.
2. R. Daiyan, I. Macgill, and R. Amal, "Opportunities and Challenges for Renewable Power-to-X," *ACS Energy Lett.*, vol. 5, no. 12, pp. 3843–3847, 2020, doi: https://doi.org/10.1021/acsenergylett.0c02249.
3. K. Adeli, M. Nachtane, A. Faik, D. Saifaoui, and A. Boulezhar, "How Green Hydrogen and Ammonia Are Revolutionizing the Future of Energy Production: A Comprehensive Review of the Latest Developments and Future Prospects," *Appl. Sci.*, vol. 13, no. 15, 2023, doi: https://doi.org/10.3390/app13158711.
4. W. Ajeeb, D. Melo, R. Costa, and P. Baptista, "Life cycle analysis of hydrotreated vegetable oils production based on green hydrogen and used cooking oils," Fuel, vol. 390, no. February, p. 134–749, 2025, doi: https://doi.org/10.1016/j.fuel.2025.134749.

5. W. Ajeeb, R. Costa, and P. Baptista, "Life cycle assessment of green hydrogen production through electrolysis : A literature review," Sustain. Energy Technol. Assessments, vol. 69, no. May, p. 103923, 2024, doi: https://doi.org/10.1016/j.seta.2024.103923.
6. T. Yusaf *et al.*, "Sustainable hydrogen energy in aviation—A narrative review," *Int. J. Hydrogen Energy*, vol. 52, pp. 1026–1045, 2024, doi: https://doi.org/10.1016/j.ijhydene.2023.02.086.
7. L. Phan-Van, V. N. Dinh, R. Felici, and T. N. Duc, "New models for feasibility assessment and electrolyser optimal sizing of hydrogen production from dedicated wind farms and solar photovoltaic farms, and case studies for Scotland and Vietnam," *Energy Convers. Manag.*, vol. 295, no. September, p. 117597, 2023, doi: https://doi.org/10.1016/j.enconman.2023.117597.
8. R. Assunção, W. Ajeeb, F. Eckl, D. M. Gomes, and R. C. Neto, "Decentralized hydrogen-oxygen co-production via electrolysis for large hospitals with integrated hydrogen refueling station," Int. J. Hydrogen Energy, vol. 103, no. January, pp. 87–98, 2025, doi: https://doi.org/10.1016/j.ijhydene.2025.01.169.
9. M. Khalil and I. Dincer, "Investigation of a community-based clean energy system holistically with renewable and hydrogen energy options for better sustainable development," *J. Clean. Prod.*, vol. 440, no. December 2023, p. 140895, 2024, doi: 10.1016/j.jclepro.2024.140895.
10. S. H. Han, Y. S. Lee, J. R. Cho, and K. H. Lee, "Efficiency analysis of air-fuel and oxy-fuel combustion in a reheating furnace," *Int. J. Heat Mass Transf.*, vol. 121, pp. 1364–1370, 2018, doi: https://doi.org/10.1016/j.ijheatmasstransfer.2017.12.110.
11. C. Chin, Z. Kamin, M. H. V. Bahrun, and A. Bono, "The Production of Industrial-Grade Oxygen from Air by Pressure Swing Adsorption," *Int. J. Chem. Eng.*, vol. 2023, 2023, doi: https://doi.org/10.1155/2023/2308227.
12. "Oxygen price outlook | Q4 2022," *Intratec Solutions*, 2023. https://medium.com/intratec-products-blog/oxygen-prices-latest-historical-data-in-several-countries-b180fad7dafa
13. G. Squadrito, A. Nicita, and G. Maggio, "A size-dependent financial evaluation of green hydrogen-oxygen co-production," *Renew. Energy*, vol. 163, pp. 2165–2177, 2021, doi: https://doi.org/10.1016/j.renene.2020.10.115.
14. R. Assunçao, F. Eckl, C. Ramos, C. Correia, and R. C. Neto, "Oxygen liquefaction economical value in the development of the hydrogen economy," *Int. J. Hydrogen Energy*, vol. 62, pp. 109–118, 2024, doi: https://doi.org/10.1016/j.ijhydene.2024.02.205.
15. Q. Hassan, A. Z. Sameen, H. M. Salman, and M. Jaszczur, "Large-scale green hydrogen production via alkaline water electrolysis using solar and wind energy," *Int. J. Hydrogen Energy*, vol. 48, no. 88, pp. 34299–34315, 2023, doi: https://doi.org/10.1016/j.ijhydene.2023.05.126.
16. Comissão Europeia, "Plano Nacional Energia E Clima 2021-2030 (Pnec 2030)," 2018.
17. Zulfhazli, A. R. Keeley, S. Takeda, and S. Managi, "A systematic review of the techno-economic assessment of various hydrogen production methods of power generation," *Front. Sustain.*, vol. 3, 2022, doi: https://doi.org/10.3389/frsus.2022.943145.
18. "International Organization for Standardization (ISO). Environmental management e life cycle assessment e principles and framework," 2006. https://www.iso.org/obp/ui/#iso:std:iso:14040:ed-2:v1:en
19. A. Lozanovski, O. Schuller, M. Faltenbacher, M. Fischer, K. Sedlbauer, and Fraunhofer-Institut für Bauphysik, *Guidance document for performing LCA on hydrogen production systems*, vol. 9. 2013.
20. European Comission, "Recommendations for Life Cycle Impact Assessment in the European context - Based on existing environmental impact assessment models and factors," 2011.
21. S. Seetharaman, R. Balaji, K. Ramya, K. S. Dhathathreyan, and M. Velan, "Graphene oxide modified non-noble metal electrode for alkaline anion exchange membrane water electrolyzers," *Int. J. Hydrogen Energy*, vol. 38, no. 35, pp. 14934–14942, 2013, doi: https://doi.org/10.1016/j.ijhydene.2013.09.033.
22. I. Abe, "Alkaline Water Electrolysis. In Energy Carriers and Conversion Systems," *UNESCO–Encyclopedia Life Support Syst. Chiba*, vol. I, 2009.
23. S. Shiva Kumar, S. U. B. Ramakrishna, S. V. Krishna, K. Srilatha, B. R. Devi, and V. Himabindu, "Synthesis of titanium (IV) oxide composite membrane for hydrogen production through alka-

line water electrolysis," *South African J. Chem. Eng.*, vol. 25, pp. 54–61, 2018, doi: https://doi.org/10.1016/j.sajce.2017.12.004.
24. W. Ajeeb, P. Baptista, and R. Costa, "Life cycle analysis of hydrogen production by different alkaline electrolyser technologies sourced with renewable energy," Energy Convers. Manag., vol. 316, p. 118840, 2024, doi: https://doi.org/10.1016/j.enconman.2024.118840.
25. W. Ajeeb and S. M. S. Murshed, "Pool boiling heat transfer characteristics of new and recycled alumina Nanofluids," Nanomaterials, vol. 13, p. 1040, 2023, doi: https://doi.org/10.3390/nano13061040.
26. W. Ajeeb and S. M. S. Murshed, "Pool boiling heat transfer characteristics of SiO 2 and BN Nanoparticles dispersed mono and hybrid Nanofluids," Nanomaterials, vol. 11, pp. 1–13, 2023e.
27. W. Ajeeb and S. M. S. Murshed, "Comparisons of numerical and experimental investigations of the thermal performance of Al2O3 and TiO2 Nanofluids in a compact plate heat exchanger," Nanomaterials, vol. 12, no. 20, p. 3634, 2022, doi: https://doi.org/10.3390/nano12203634.
28. W. Ajeeb, D. Roque, and S. M. S. Murshed, "Thermal-hydraulic performance of Al2O3 Nanofluids under laminar flow in a Minitube," Int. J. Energy a Clean Environ., vol. 23, no. 5, pp. 39–59, 2022, doi: https://doi.org/10.1615/InterJEnerCleanEnv.2022041476.
29. W. Ajeeb and S. M. S. Murshed, "Characterization of Thermophysical and electrical properties of SiC and BN Nanofluids," Energies, vol. 16, no. 9, 2023, doi: https://doi.org/10.3390/en16093768.
30. "European Communities: Wind Turbine Grid Connection and Interaction," 2001. [Online]. Available: https://energy.ec.europa.eu/publications/wind-turbine-grid-connection-and-interaction_en
31. Eurelectric, "Composition of Portugal's electricity generation (2023)," 2023. https://electricity-data.eurelectric.org/electricity-generation/electricity-generation-by-fuel.html
32. Eurelectric, "Composition of Spain's electricity generation (2023)," 2023. https://electricity-data.eurelectric.org/electricity-generation/electricity-generation-by-fuel.html
33. A. Lotrič, M. Sekavčnik, I. Kuštrin, and M. Mori, "Life-cycle assessment of hydrogen technologies with the focus on EU critical raw materials and end-of-life strategies," *Int. J. Hydrogen Energy*, vol. 46, no. 16, pp. 10143–10160, 2021, doi: https://doi.org/10.1016/j.ijhydene.2020.06.190.
34. S. Rajeshwar, Krishnan; McConnell, Robert; Licht, *Towards a Renewable energy future : Solar Hydrogen Generation*, vol. 53, no. 9. 2019.
35. J. Sousa *et al.*, "Renewable energy communities optimal design supported by an optimization model for investment in PV/wind capacity and renewable electricity sharing," *Energy*, vol. 283, no. January, 2023, doi: https://doi.org/10.1016/j.energy.2023.128464.
36. X. Zhang, C. Bauer, C. L. Mutel, and K. Volkart, "Life Cycle Assessment of Power-to-Gas: Approaches, system variations and their environmental implications," *Appl. Energy*, vol. 190, pp. 326–338, 2017, doi: https://doi.org/10.1016/j.apenergy.2016.12.098.
37. L. Schleisner, "Life cycle assessment of a wind farm and related externalities," *Renew. Energy*, vol. 20, no. 3, pp. 279–288, 2000, doi: https://doi.org/10.1016/S0960-1481(99)00123-8.
38. D. H. N. C. Reichel, A. Müller, L. Friedrich, S. Herceg, M. Mittag, A. Protti, "CO2 Emissions of Silicon Photovoltaic Modules—Impact of Module Design and Production Location," *8th World Conf. Photovolt. Energy Convers.*, no. September, pp. 1617–1619, 2022.
39. S. Ghandehariun and A. Kumar, "Life cycle assessment of wind-based hydrogen production in Western Canada," *Int. J. Hydrogen Energy*, vol. 41, no. 22, pp. 9696–9704, 2016, doi: https://doi.org/10.1016/j.ijhydene.2016.04.077.
40. S. Sadeghi and S. Ghandehariun, "Environmental impacts of a standalone solar water splitting system for sustainable hydrogen production: A life cycle assessment," *Int. J. Hydrogen Energy*, vol. 48, no. 50, pp. 19326–19339, 2023, doi: https://doi.org/10.1016/j.ijhydene.2023.01.234.
41. M. Tariq, A. Siddhantakar, J. D. Sherman, A. Cimprich, and S. B. Young, "Life cycle assessment of medical oxygen," *J. Clean. Prod.*, vol. 444, no. October 2023, p. 141126, 2024, doi: https://doi.org/10.1016/j.jclepro.2024.141126s.

Part III
Energy Efficiency and Energy Management

Assessment of Renewable Energy Efficiency and Its Impact on Sustainable Development: A Non-parametric Approach Based on Evidence from Developed and Emerging Economies

Nirupam Mukhopadhyay and Narayan Chandra Nayak

Abstract It is evident that developed economies have made significant progress in terms of renewable energy due to their early advancements. In contrast, emerging economies are falling behind in this aspect. While there has been some catching up by emerging economies in the last decade, there is still a considerable distance for them to cover. Our analysis mainly focuses on calculation of renewable energy efficiency in developed as well as in emerging economies using Stochastic Frontier Analysis or SFA, which supports the above conclusions and underscores the need for further developments. This work also introduces a non-parametric model to check for the relationship between sustainability and renewable energy efficiency. How this efficiency impacts sustainability through non-parametric estimation known as Local Linear Dummy Variable Estimation, and obtained positive results indicating a pressing need for the development of renewable energy—a contribution that will lead towards greater sustainability efforts.

Keywords Renewable energy efficiency · Sustainability · Stochastic Frontier Analysis or SFA · Non-parametric estimation · Local linear dummy variable estimation

1 Introduction

Renewable energy has become a crucial component in the pursuit of sustainable development, with its potential to mitigate climate change and reduce dependence on fossil fuels. There is a need for renewable energy sources such as solar, wind, hydropower, and biomass that have continuous improvement in terms of efficiency

N. Mukhopadhyay (✉) · N. C. Nayak
Department of Humanities and Social Sciences, Indian Institute of Technology, Kharagpur, India

Z. Peng (ed.), *Technological Advancements and Future Directions in Green Energy*, Green Energy and Technology,
https://doi.org/10.1007/978-3-031-83203-1_8

and cost-effectiveness. These advancements in renewable energy technologies have allowed for greater integration of clean energy into the energy mix, resulting in reduced greenhouse gas emissions and improved air quality for developed economies [3]. Moreover, the adoption of renewable energy systems has shown positive impacts on economic growth and job creation, particularly in beachfront and remote areas where ocean energy innovations can empower energy security and contribute to economic growth. By analyzing the latest advancements in renewable energy and their implications for sustainability, economic growth, job creation, and energy security, this work seeks to provide valuable insights into the potential of renewable energy to drive sustainable development [2, 4]. The role of renewable energy in reducing carbon emissions and transitioning away from traditional fossil fuels has become increasingly vital. There is a need to investigate deeper into the efficiency of renewable energy sources such as solar, wind, hydropower, and biomass, and to evaluate their impact on sustainable development in both developed and emerging economies. So that proper policies can be taken to bolster the rate of sustainable development [10, 14]. The current status of renewable energy efficiency varies across economies, with developed countries often leading in technological advancements and deployment of renewable energy systems [10, 21]. Solar technology, for example, has witnessed remarkable improvements in efficiency and cost reduction, making it more competitive with conventional energy sources in many regions [26]. Similarly, wind power technologies have become more efficient and widespread, contributing significantly to the renewable energy capacity of several developed nations [20, 22].

In comparison, emerging economies are also making substantial progress in adopting renewable energy technologies [11, 13]. Rapid advancements in solar panel manufacturing, for instance, have led to increased affordability and accessibility of solar energy in many developing countries. Furthermore, innovative financing mechanisms and policy incentives have facilitated the expansion of renewable energy infrastructure in these regions [25].

The impact of renewable energy on sustainable development is multi-faceted and extends beyond environmental benefits. In developed economies, the increasing deployment of renewable energy systems has contributed to job creation in the clean energy sector, while also enhancing energy security and reducing reliance on imported fossil fuels ([25]). Moreover, the integration of renewable energy into the energy mix has played a vital role in diversifying the energy sources, thereby improving overall energy resilience. In emerging economies, the adoption of renewable energy has not only bolstered energy security but has also brought electricity access to remote and underserved communities. This has led to socio-economic development, enabling opportunities for education, healthcare, and other essential services that are dependent on reliable energy access. Additionally, the expansion of renewable energy infrastructure has stimulated economic growth through investment opportunities and the establishment of local manufacturing and assembly facilities for renewable energy components [11, 22]. The assessment of renewable energy efficiency and its impact on sustainable development presents a compelling case for continued investments and policy support to accelerate the global transition towards renewable energy [8]. As the world strives to achieve sustainable development goals and address the challenges of climate change, renewable energy stands as a key enabler for a cleaner, more resilient, and inclusive energy future.

Despite significant progress in renewable energy technologies and their positive impact on sustainable development, several barriers continue to impede widespread deployment. One of the major challenges is the intermittency of renewable energy sources such as solar and wind, which necessitates the development of efficient storage solutions and smart renewable energy infrastructure to ensure reliable energy supply. In summary, we can conclude that energy efficiency is a critical concern for both the current and future generations. Improving performance in sustainable development goals and overall sustainability hinges on assessing its impact alongside other factors. To accomplish this, we are utilizing SFA to calculate renewable energy efficiency and employing non-parametric models to estimate its effects on sustainability. This work is organized as follows: First, there is a brief literature review on calculating renewable energy efficiency. Then we will delve into the details of the data and sources. Following that, we will discuss the results before concluding with references.

2 Literature Review

Energy efficiency is often studied using Data Envelopment Analysis (DEA) or Stochastic Frontier Analysis (SFA), DEA does not require summing different outputs and has zero model misspecification possibility. Literature focusing on energy efficiency commonly uses the SFA method, as demonstrated by various studies analyzing energy efficiency in different sectors and countries, for example, studies were conducted in Canadian industries and in Japanese power plants. Examined energy utilization efficiency and CO_2 reduction potentials in the greenhouse cucumber industry in Iran using a similar study. Chang et al. [9] applied both DEA and ANN to analyze the transportation sector in China, finding that environmental efficiency was less than 50% and CO_2 reduction potential varied between 1.6 million tons and 33 million tons. Researcher also uses SFA to evaluate efficiency due to its consideration of stochastic effect. Boyd [6] and Boyd et al. [7] used SFA to estimate energy efficiency in the US manufacturing sector at the plant level. Shi et al. [23] measured the impacts of main factors on energy efficiency in China between 1980 and 2005 by using variance decomposition method and found that total factor productivity contributed 36%, capital-energy ratio 45%, and labor-energy ratio 18%. Alfonso et al. [1] used SFA to study the energy efficiency of various sectors in Spain, showing each sector has a potential for 20% energy conservation. Massimo and Lester collected data from 48 US states from 1995–2007, analyzed residential sector's energy consumption and efficiency, finding that while energy intensity could not fully represent, energy efficiency was properly measured by SFA model. Lin and Yang [18] analyzed the energy efficiency of the thermal power industry in China from 2005 to 2010, finding an average energy efficiency of 0.85 and a cumulative energy conservation potential of 551 Mtce. Chen et al. [9] examined fossil-fuelled power plants in China between 1999 and 2011, identifying significant variations in efficiency based on size, location, ownership, and electricity generation sources. Katuwal et al. [15] measured the effectiveness of suppression resource inputs for containing large wildfires using SFA, concluding that bulldozers and fire engines had more impact than firefighter crews while emphasizing the greater efficiency when combined with natural or man-made breaks within previously burned areas by wildfire.

Most of the existing literature on the efficiency of REG focuses on calculating and analyzing factors influencing solar, wind, geothermal, and bioenergy generation efficiency at a technical level. Scholars emphasize maximizing utilization in solar power by analyzing energy loss within photovoltaic modules, converters, inverters, and booster transformers to improve overall power generation efficiency. Factors such as weather conditions and plant-specific situations also play a significant role in determining solar power generation efficiency. Research studies analyze factors such as solar radiation, photovoltaic module characteristics, and maximum power peak tracking to improve battery photoelectric conversion efficiency, optimize sunlight tracking, enhance grid-connected technology for solar generation. For wind generation, the focus is on gradient estimation of the system and optimal torque maximum power point tracking using an optimal torque compensator. Geothermal power research explores impacts of condensate temperature and mass concentration of ammonia on geothermal systematic cycle measurements. There is a positive correlation between system cycle efficiency and ammonia mass concentration, influencing economic decline. Research on bio-energy generation includes evaluation, factor analysis, and improvement measures. Regional differences in biomass energy distribution exist in China with high efficiency in the south, southwest, and east but with room for improvement in the north. Existing studies mainly focus on maximizing REG efficiency at the technical level, with little consideration given to the economic aspect. This paper takes an economic approach and explores the impact of government subsidy policies, which include financial support, tax rebates, and non-monetary assets.

3 Data and Methodology

3.1 Data

We use yearly data concerning 41 countries both developed and emerging economies from MSCI list of country classification. Our data cover the period 1999–2022. Thus, there are 40 cross-section units and 24 time-series observations per cross-section unit. One of our main variable is renewable energy efficiency or reneweffi. To calculate reneweffi we have used Stochastic Frontier Analysis or SFA as describes step by step in methodology part. For other variable and details please see Tables 1 and 2.

Table 1 Variable construction and details

Variable name	Details	Source
Sustainability sus	Value of 17 SDG or sustainable development goals. Using those 17 values we have constructed an index using principle component analysis or PCA	Own calculation and SDG data from UNDP
Renewable energy efficiency reneweffi	Calculated using SFA	Own calculation
Renewable energy index renewindex	Index using all the renewable energy sources data	Own calculation
Research and development expenditure r&d	% of GDP	World Bank
Human capital hc	Measures which countries are best in mobilizing their human capital	World Bank
Gross fixed capital formation gfcf	% of GDP	World Bank
Hydro	Hydro power capacity	IEA
Wind	Wind capacity/GW	IEA
Solar	Solar capacity/GW	IEA
Biogas	Biofuel production/Mt	IEA
CO2 emission	Carbon dioxide emission/LN	World Bank
Popu	Population density	World Bank
Urab	Urbanization rate/%	World Bank

Source: By author

Table 2 Descriptive statistics

Variable	Obs	Mean	Std. Dev.	Min	Max
sus	960	72.88	7.06	51.66	86.76
reneweffi	960	0.41	0.11	0.23	0.77
renewindex	960	0.21	2.02	−0.82	26.75
rd	960	1.53	1.08	−0.19	5.70
hc	960	2.98	0.51	1.64	3.89
gfcf	960	3.32	8.85	−40.4	100.69
hydro	960	0.711	1.53	0	12.49
wind	960	0.14	0.50	0	7.15
solar	960	0.057	0.24	−0.01	4.01
biogeo	960	0.106	.201	0	2.029
co2 emission	960	7.209	4.22	0.72	20.47
popu	960	141.52	138.66	2.44	531.10
uarb	960	73.25	14.50	27.45	98.15

Source: Authors calculation

3.2 Methodology

Stochastic Frontier Analysis or SFA to evaluate efficiency because it takes stochastic noise into account. This method's discriminating power allows it to effectively evaluate energy efficiency performance. We estimate the production frontier (maximum output observed for given inputs) and then integrate the exogenous variables that affect performance or technical inefficiency. In a stochastic frontier model, the error term has a one-sided component and a traditional random error. The one-sided component can measure the effects of technical inefficiency, and the traditional random error can determine the effects of measurement error [16]. The production function is shown as follows

$$y_{it} = \alpha + \beta x_{it} + v_{it} - u_{it} \tag{1}$$

y_{it} is the output; x_{it} is the set of inputs; v_{it} and u_{it} are random error and one sided efficiency error term.

There are two ways to determine efficiency in SFA model. Two-step model, first we estimate inefficiencies without considering all the exogenous impacting factors then go for regression of the inefficiency estimate with the dependent variable. But if there is a correlation amongst the two, then this model becomes inefficient. In one step (Table 3 and 4):

One-step model estimates associations in the production as well as the inefficiency factors. The inefficiency effect apparatuses of the SFA contain the explanatory variables that affect technical inefficiency (TIE).

$$u_{it} = \gamma z_{it} + w_{it}; u_{it} > 0; \tag{2}$$

$$u_{it} \sim i.i.d.\infty N\left(\gamma z_{it}, \sigma_u^2\right) \tag{3}$$

$$w_{it} \sim i.i.d.\infty N\left(-\gamma z_{it}, \sigma_u^2\right) \tag{4}$$

$$v_{it} \sim i.i.d.\infty \; N\left(0, \sigma_v^2\right) \tag{5}$$

where z_{it} is a vector of explanatory variables that affect TIE, γ is vector of inefficiency variables. w_{it} describes random components of inefficiency. From combing Eqs. (1) and (2–5) we will deduce our SFA model, i.e.,

$$y_{it} = \alpha + \beta x_{it} + v_{it} - u_{it}\left(z_{it}, \gamma\right) \tag{6}$$

In Eq. (4), $y_{it}$is the output variable which is total energy consumed by i^{th} country at t^{th} time. x_{it} is the set of inputs, i.e., solar, wind, thermal, and biofuel renewable energy sources as inputs along with capital and labour. This input output set up in modelled by Cobb-Douglas Production Function (CDPF) Eq. (7).

Table 3 Correlation table

	(1)	(2)	(3)	(4)	(5)	(6)	(7)	(8)	(9)	(10)	(11)	(12)	(13)
(1) sus	1												
(2) reneweffi	−0.134	1											
(3) renewindex	0.102	0.003	1										
(4) r&d	0.607	−0.108	0.242	1									
(5) hc	0.746	−0.048	0.224	0.707	1								
(6) gcfc	−0.133	0.61	−0.041	−0.105	−0.094	1							
(7) hydro	−0.053	−0.067	0.538	0.034	0.073	−0.002	1						
(8) wind	0.123	0.013	0.915	0.212	0.209	−0.032	0.296	1					
(9) solar	0.118	0.02	0.807	0.234	0.19	−0.035	0.17	0.797	1				
(10) biogeo	0.102	0.016	0.887	0.258	0.229	−0.053	0.5	0.698	0.565	1			
(11) CO_2	0.384	0.006	0.201	0.506	0.653	−0.061	0.237	0.155	0.077	0.215	1		
(12) popu	−0.007	0.009	−0.022	0.394	0.1	−0.016	−0.278	−0.025	0.115	0.044	−0.05	1	
(13) uarb	0.544	−0.16	0.116	0.493	0.594	−0.106	0.108	0.08	0.085	0.109	0.428	−0.024	1

Table 4 Renewable energy efficiency trends for developed and emerging economies

(a) Renewable energy efficiency developed economies	(b) Renewable energy efficiency emerging economies.
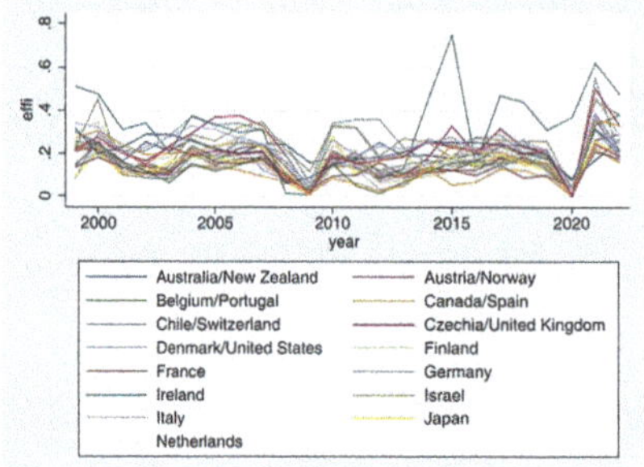	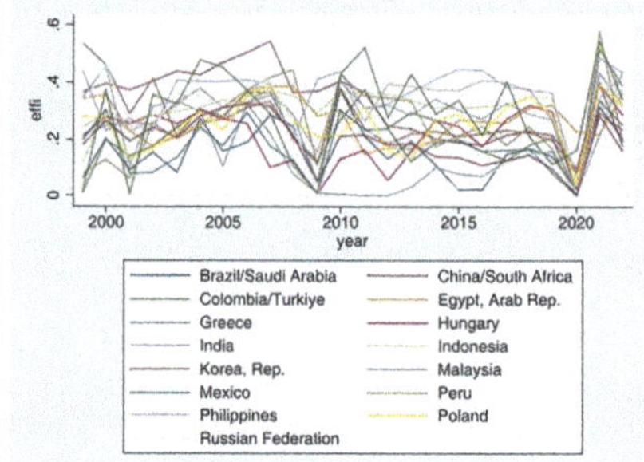

Source: Authors calculation

$$\mathrm{Lny}_{it} = \beta_0 + \sum_{i=1}^{6}\sum_{t=1}^{n}\left[\beta_{it}x_{it}\right] + v_{it} - u_{it}\left(z_{it},\gamma\right) \tag{7}$$

If we simplify Eq. (7), we get

$$\begin{aligned}\mathrm{Lny}_{it} = \beta_0 + \beta_1\mathrm{solar}_{it} + \beta_2\mathrm{wind}_{it} + \beta_3\mathrm{tharmal}_{it} + \\ \beta_4\mathrm{biofuel}_{it} + \beta_5\mathrm{labour}_{it} + \beta_6\mathrm{capital}_{it} + v_{it} - u_{it}\left(z_{it},\gamma\right)\end{aligned} \tag{8}$$

And u_{it} the inefficiency parameter is a function of CO_2 emission *(CO2)*, population *(popu)* and urbanization *(urab)* this will affect the renewable energy efficiency.

$$u_{it} = \gamma_0 + \gamma_1\left(\mathrm{CO}_{2it}\right) + \gamma_2\left(\mathrm{popu}_{it}\right) + \gamma_3\left(\mathrm{urab}_{it}\right) + w_{it} \tag{9}$$

Technical Efficiency (TEF) can be calculated transforming Eq. (8),

$$\mathrm{TEF}_{it} = f\left(I_{it},;O_{it},;Z_{it},;\beta\right)e^{v_{it}+u_{it}} \tag{10}$$

In Eq. (10), *f*(.) demand function in a frontier. I_{it} *is the set of inputs*; O_{it}represents the output factor of y; Z_{it} represents the external factors; SFA model has mixed error ($v_{it} + u_{it}$), inefficiency can be separated.

$$\mathrm{EFF}_{it} = \frac{E\left((E_{it}|u_{it} = 0, K_{it}, L_{it}, O_{it}, Z_{it}\right)}{E(TE_{it}|u_{it} \neq 0, K_{it}, L_{it}, O_{it}, Z_{it})} = \exp\left(-u_{it}\right) \tag{11}$$

In Eq. (11), *E*(.) is a conditional expectation and EFF_{it} is efficiency vector (be calculated) [5, 16, 19].

After we calculate the efficiency, we will see how that efficiency affects the overall sustainability. To analyze the impact of efficiency along with other variables on sustainability, we will use non-parametric model. We are giving a short note on non-parametric estimation technique. We will use local Linear Dummy Variable Estimation or LLDVE to estimate the time varying trend and coefficient [12, 17, 24, 27].

Parametric panel data model in our case,

$$\text{sus}_{it} = \alpha_i + \beta_1 \text{EFF}_{it} + \beta_2 rd_{it} + \beta_3 hc_{it} + \beta_4 gfcf_{it} + e_{it} \tag{12}$$

For the above parametric model, we will test for Cross Sectional Dependence or CSD, unit root, and stationary test. We will study the above parametric model for overall sample, developed and emerging economies.

In case of non-parametric model Eq. (12) looks like,

$$\text{sus}_{it} = f_t + X_{it}^T \beta_t + \alpha_i + e_{it} i = 1,2,3,\ldots,N; t = 1,2,\ldots T \tag{13}$$

$$X_{it} = \left(\text{EFF}_{it}, rd_{it}, hc_{it}, gfcf_{it}\right) \tag{14}$$

$$\beta_t = \left(\beta_{t,1}, \beta_{t,2}, \beta_{t,3}, \beta_{t,4}\right), \tag{15}$$

In above equation, $f_t = f(t/T)$ is the unknown trend; $\beta_{t,j}$ and f_t time varying coefficient and trend function. Our objective is to find out the relationship between α_i and e_{it}. We assume,

$\sum_{i=1}^{N} \alpha_i = 0.$

Now we transform Eq. (13) as,

$$\widetilde{\text{sus}} = \tilde{f} + B\widetilde{(X,\beta)} + \tilde{D}\alpha + \tilde{e} \tag{16}$$

~~Here~~

$\text{sus} = \left(Y_1^T, \ldots\ldots Y_n^T\right)^T$, $\text{sus}_i = (Y_{i,1} \ldots\ldots Y_{iT})^T$, $\tilde{e} = \left(e_1^T, \ldots\ldots e_n^T\right)^T$, $e_i = (e_{i1} \ldots\ldots e_{iT})^T$

$$\alpha^* = \left(\alpha_1, \ldots \alpha_N\right)^T$$

$$\tilde{D} = I_N \otimes \overline{I_N}$$

I_N is a k*1 vector of ones.

Now we can write Eq. (16) as,

$$\widetilde{\text{sus}} = \tilde{f} + B\widetilde{(X,\beta)} + \widetilde{D^*}\alpha^* + \tilde{e} \tag{17}$$

Here

$$D^* = \left(-I_{N-1}, I_{N-1} \right)^T \otimes I_N \tag{18}$$

We will use the above prescribe methodology for our non-parametric analysis.

4 Result and Discussion

First, we will discuss the renewable energy efficiency results. Using SFA, we have calculated the renewable energy efficiency Table 5. We have year wise calculated the efficiency value of all the inputs used. Along with that, we have also checked the significance level of the input and inefficiency variables in Table 6. Table 5 shows that CO2 emission has significant negative relation with renewable energy efficiency. So counties with more carbon emission should reduce their dependency on carbon emission technologies. They should go more towards renewable energy solutions. Urbanization (urab) increases energy demand as well as consumption. In our case, we observe that urbanization is positively contributing towards renewable energy efficiency or reEFFI.

Our four renewable energy components are hydro, solar, wind, and biogas and final output is total energy outcome. Here also we found that all the aforementioned components are positively contributing towards the renewable energy efficiency. We have also reported the renewable energy efficiency performance for all our selected countries in Table 6. It is observed that developed economies are performing better in terms of efficiency compared to the emerging economies.

Now we will see how energy efficiency impact sustainability. Table 6 represents the parametric (fixed effect) model results. To check the impact of renewable energy, we have used both renewable energy efficiency as an index of renewable energy and found that both are significantly impacting overall sustainability. We carried out this analysis for developed as well as emerging economies and found significant positive results.

As previously noted, the use of a non-parametric time-varying coefficients panel data analysis with fixed effects allows for an analysis that does not restrict the relationship between the variables to a predefined functional shape. Moreover, nonparametric estimators evaluate how the response of the dependent variable adapts to incremental variations in the independent variables. The main goal of this section is to examine how the responsiveness of renewable energy efficiency and additional variables evolves over time. We utilize a bivariate nonparametric method to evaluate the hypothesis of environmental improvement across 41 developed and developing countries. Contrary to this, our findings suggest a positive trend in renewable energy efficiency from 1999 to 2022, contributing to overall sustainability.

The bivariate nonparametric analysis, using renewable energy efficiency as the predictor, corroborates the results from the previous subsection's parametric regressions. A one percent rise in renewable energy efficiency corresponds to an increase

Table 5 Estimation results of SFA

Variable name	Coefficient	Std. error	P-value
Input			
Hydro	0.44***	0.09	0.00
Wind	0.56**	0.57	0.03
Solar	0.52*	0.01	0.06
Biogas	3.27**	0.90	0.00
Determinants of inefficiency			
CO2 emission	−0.07***	0.02	0.00
Popu	0.01**	0.0007	0.06
Urab	0.75***	0.11	0.00

Source: Authors calculation
***, **&* are 1%, 5% and 10% level of significance respectively

Table 6 Fixed effect estimation

	Overall		Developed		Emerging	
Reneweffi	0.39*** (0.52)		0.06** (0.04)		1.14* (1.03)	
Index_renew		0.49*** (0.06)		0.28*** (0.50)		1.12*** (0.23)
Research and development r&d	1.11*** (0.16)	0.93*** (0.16)	1.34*** (0.15)	1.16*** (0.15)	0.23* (0.11)	0.45* (0.32)
Human capital	10.45*** (0.35)	9.69*** (0.35)	12.85*** (0.44)	12.30*** (0.44)	9.95*** (0.54)	8.12*** (0.65)
GFCF	−0.01 (−0.006)	−0.007 (0.004)	0.004 (0.005)	−0.004	−0.02 (0.01)	−0.01

Source: Authors calculation
***, **&* are 1%, 5% and 10% level of significance respectively

in sustainability. Figures 1a, 2a, and 3a depict the dynamics of this association. Our research uncovers a nonlinear link between energy efficiency and sustainability, with developed economies exhibiting both positive and negative trends due to their existing proficiency in renewable energy use, while emerging markets display an unbroken positive trajectory.

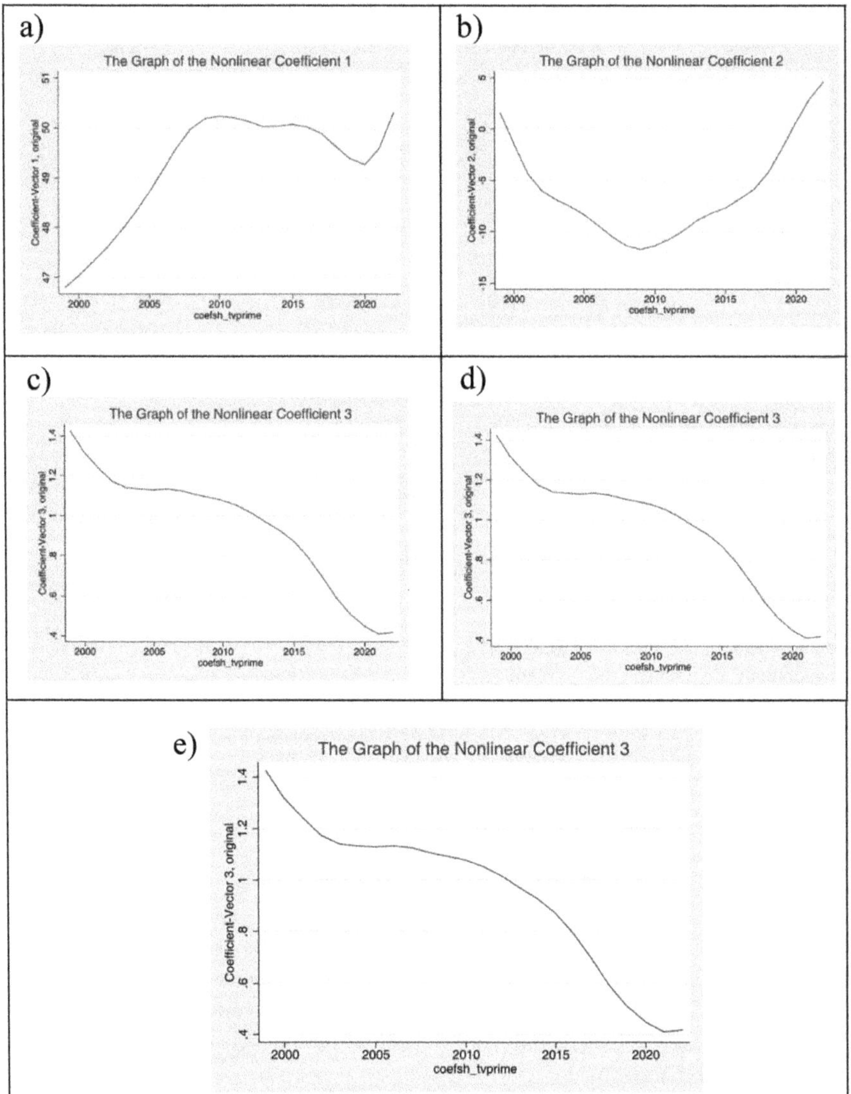

Fig. 1 Time varying response of efficiency to sustainability along with other variable: overall economies. Local linear dummy variable estimates of common trend and coefficients. (**a**) trend of intercept term; (**b**) coefficient trend of efficiency; (**c**) coefficient trend of r&d; (**d**) coefficient trend of human capital; (**e**) coefficient trend of gfcf. (Source: Authors calculation)

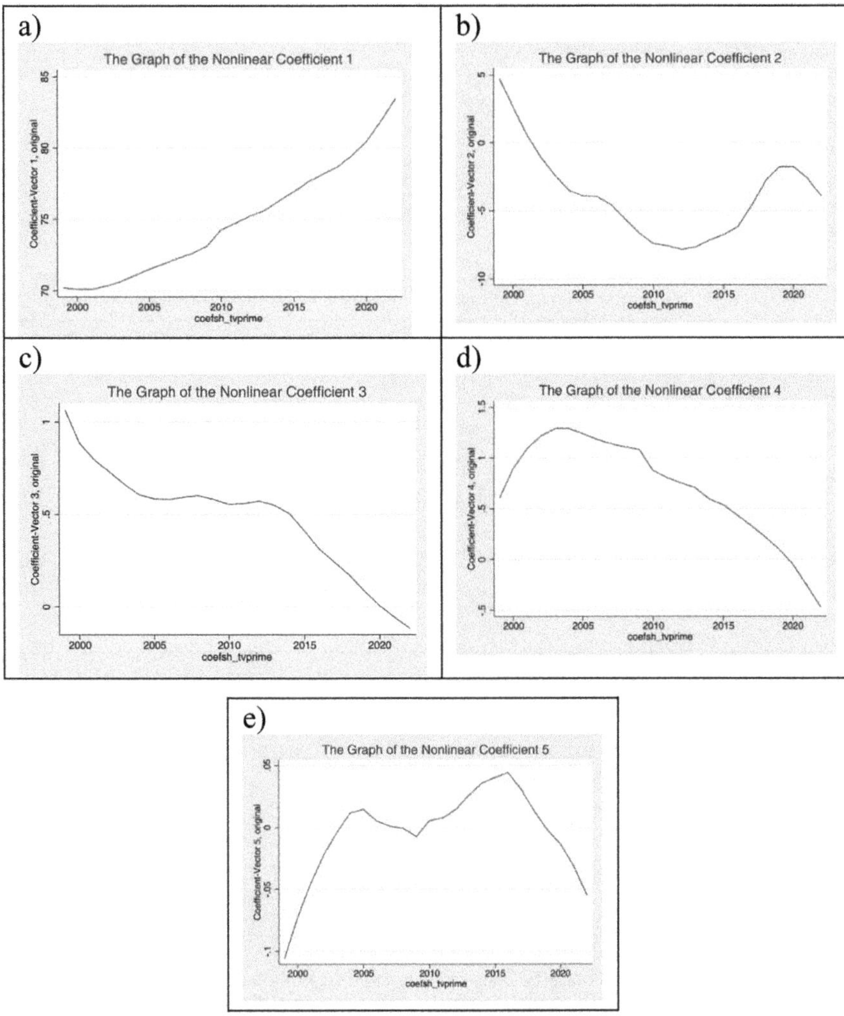

Fig. 2 Time varying response of efficiency to sustainability along with other variable: developed economies. Local linear dummy variable estimates of common trend and coefficients. (**a**) trend of intercept term; (**b**) coefficient trend of efficiency; (**c**) coefficient trend of r&d; (**d**) coefficient trend of human capital; (**e**) coefficient trend of gfcf. (Source: Authors calculation)

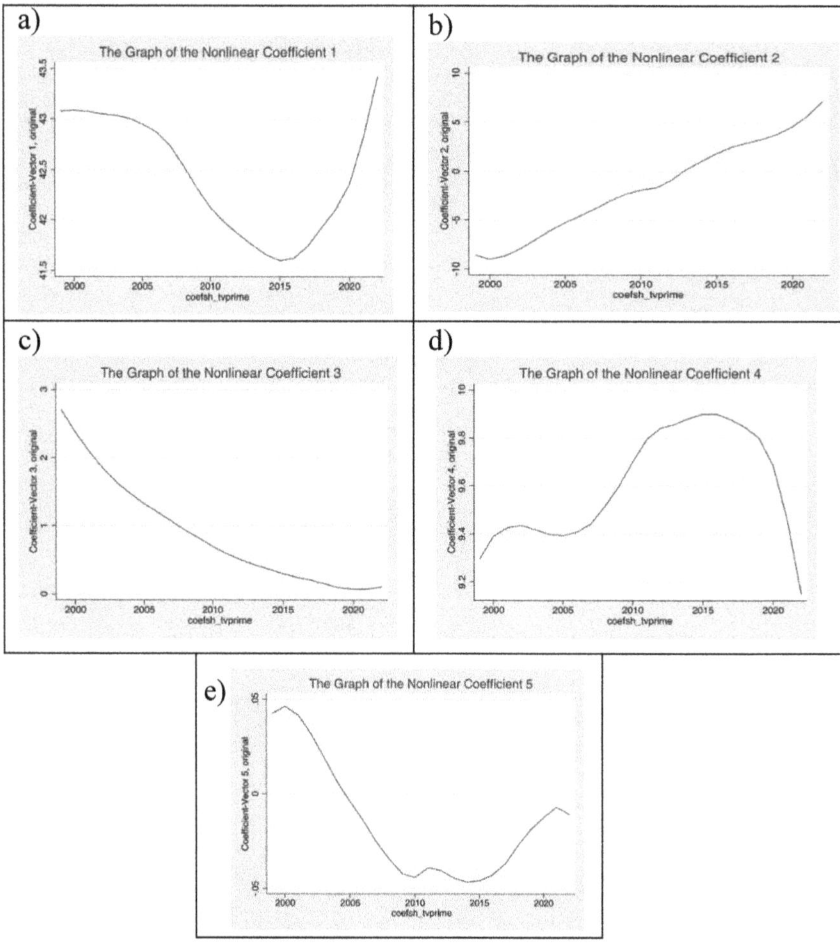

Fig. 3 Time varying response of efficiency to sustainability along with other variable: emerging economies. Local linear dummy variable estimates of common trend and coefficients. (**a**) trend of intercept term; (**b**) coefficient trend of efficiency; (**c**) coefficient trend of r&d; (**d**) coefficient trend of human capital; (**e**) coefficient trend of gfcf. (Source: Authors calculation)

5 Conclusion

From the above analysis, we found that developed economies are much more efficient in terms of renewable energy due to their early progress. On the other hand, emerging economies are lagging behind. Although in the last decade, emerging economies have caught up to some extent, there is still a long way to go. Our energy efficiency parameter calculated through SFA supports these findings. We then analyzed how this efficiency affects sustainability through non-parametric estimation (LLDVE) and found positive results which suggest a high need for the development of renewable energy—a contribution that will lead towards greater sustainability efforts [1, 9].

In conclusion, the analysis highlights the importance of energy efficiency and renewable energy in achieving sustainability goals. Furthermore, it emphasizes the need for continued investment and development in renewable energy technologies, especially in emerging economies. In conclusion, the analysis highlights the importance of energy efficiency and renewable energy in achieving sustainability goals. Furthermore, it emphasizes the need for continued investment and development in renewable energy technologies, particularly in emerging economies, to further contribute towards sustainability.

Looking ahead, the future of renewable energy appears promising, driven by ongoing technological innovation and favorable market dynamics. Continued research and development efforts aimed at enhancing the efficiency and reliability of renewable energy systems will be critical in further advancing the global renewable energy transition. Additionally, collaboration between developed and emerging economies in sharing best practices and knowledge transfer can accelerate the adoption of renewable energy on a global scale. In conclusion, the assessment of renewable energy efficiency and its impact on sustainable development underscores the need for holistic approaches that address both technological and socio-economic aspects. By overcoming existing barriers and seizing opportunities for international cooperation, the transition towards a cleaner and more sustainable energy landscape can be expedited, benefitting both present and future generations.

References

1. Alfonso, A., German, F., Mainar-Toledo, M. D., and Sabina, S., “Energy consumption analysis of Spanish food and drink, textile, chemical and non-metallic mineral products sectors,” Energy 42(1), 477–485 (2012).
2. Anderson S, Newell R. 2004. Information programs for technology adoption: the case of energy efficiency audits. Resour. Energy Econ. 26:27–50.
3. Asif M, Muneer T. Energy supply, its demand and security issues for developed and emerging economies. Renew Sustain Energy Rev 2007;11:1388–413.
4. Auffhammer M, Blumstein C, Fowlie M. 2008. Demand-side management and energy efficiency revisited. Energy J. 29:91–104.
5. Battese, G. and Coelli, T., “A model for technical inefficiency effects in a stochastic frontier production function for panel data,” Empirical Econ. 20, 325–332 (1995).
6. Boyd, G. A., “Estimating plant level energy efficiency with a stochastic frontier,” Energy J. 29(2), 23–43 (2008).
7. Boyd, G., Dutrow, E., and Tunnessen, W., “The evolution of the ENERGY STAR (R) energy performance indicator for benchmarking industrial plant manufacturing energy use,” J. Cleaner Prod. 16(6), 709–715 (2008).
8. Cantarero, Maria Mercedes Vanegas. “Of renewable energy, energy democracy, and sustainable development: A roadmap to accelerate the energy transition in developing countries.” Energy Research & Social Science 70 (2020): 101716.
9. Chang, Y. T., Zhang, N., Danao, D., and Zhang, N., “Environmental efficiency analysis of transportation system in China: A non-radial DEA approach,” Energy Policy 58, 277–283 (2013).

10. Connolly D, Lund H, Mathiesen BV. Smart Energy Europe: the technical and economic impact of one potential 100% renewable energy scenario for the Europe Union. Renew Sustain Energy Rev 2016;60:1634e53.
11. Florea, Nicoleta Mihaela, et al. "Linking public finances' performance to renewable-energy consumption in emerging economies of the European Union." Sustainability 13.11 (2021): 6344.
12. Hailemariam, A., Smyth, R., Zhang, X., 2019. Oil prices and economic policy uncertainty: Evidence from a nonparametric panel data model. Energy Econ. 83, 40–51.
13. Halkos, George E., and Nickolaos G. Tzeremes. "The effect of electricity consumption from renewable sources on countries' economic growth levels: Evidence from advanced, emerging and developing economies." Renewable and Sustainable Energy Reviews 39 (2014): 166–173.
14. Hanergy, Annual Development Report on Global New Energy (2016) (Hanergy, Beijing, 2016).
15. Katuwal, H., Calkin, D. E., and Hand, M. S., "Production and efficiency of large wildland fire suppression effort: A stochastic frontier analysis," J. Environ. Manage. 166, 227–236 (2016).
16. Kumbhakar, S. C. and Lovell, C. A. K., Stochastic Frontier Analysis (Cambridge University Press, Cambridge, 2003).
17. Li, D., Chen, J., Gao, J., 2011. Non-parametric time-varying coefficient panel data models with fixed effects. Econom. J. 14 (3), 387–408.
18. Lin, B. and Yang, L., "The potential estimation and factor analysis of China's energy conservation on thermal power industry," Energy Policy 62, 354–362 (2013).
19. Liu, Y. Y. and Myers, R., "Model selection in stochastic frontier analysis with an application to maize production in Kenya," J. Prod. Anal. 31(1), 33–46 (2009).
20. Lugo-Laguna, Daniel, Angel Arcos-Vargas, and Fernando Nuñez-Hernandez. "A European assessment of the solar energy cost: key factors and optimal technology." Sustainability 13.6 (2021): 3238.
21. Ma, X. H., "Ratio of renewable energy power generation in German beating the record," http://env/people.com.cn/n/2015/0805/c1010-27415137.html (2015).
22. Ortega, Margarita, et al. "Analysing the influence of trade, technology learning and policy on the employment prospects of wind and solar energy deployment: The EU case." Renewable and Sustainable Energy Reviews 122 (2020): 109657.
23. Shi, G.-M., Bi, J., and Wang, J.-N., "Chinese regional industrial energy efficiency evaluation based on a DEA model of fixing non-energy inputs," Energy Policy 38(10), 6172–6179 (2010).
24. Silvapulle, P., Smyth, R., Zhang, X., Fenech, J.-P., 2017. Nonparametric panel data model for crude oil and stock market prices in net oil importing countries. Energy Econ. 67, 255–267.
25. Tiwari, Sunil, et al. "Sustainable pathways for attaining net-zero emissions in European emerging countries—the nexus between renewable energy sources and ecological footprint." Environmental Science and Pollution Research 30.48 (2023): 105999–106014.
26. Viebahn, Peter, Yolanda Lechon, and Franz Trieb. "The potential role of concentrated solar power (CSP) in Africa and Europe—A dynamic assessment of technology development, cost development and life cycle inventories until 2050." Energy Policy 39.8 (2011): 4420–4430.
27. Zhang, Y., Su, L., Phillips, P.C.B., 2012. Testing for common trends in semi-parametric panel data models with fixed effects. Econom. J. 15 (1), 56–100.

High-Performance Passive Envelopes: A Practical Insight into the Challenges and Barriers

Ali Fathalizadeh and Ehsan Sharifi

Abstract The concept of sustainable green technologies in construction industry has drawn considerable attention among researchers in recent decades. Considering the significant role of passive cooling technologies in enhancing the thermal performance of residential buildings, this study explores the challenges and barriers towards the development of high-performance building envelopes incorporating passive technologies in Australia. This article is a part of an ongoing comprehensive research aiming at facilitating the adoption of high-performance passive envelopes in Australian residential sector. Through an in-depth investigation of the literature followed by interviews with construction professionals, the major barriers and their significance have been revealed. While the existing literature puts emphasis on the technological aspects of passive cooling, the results of this study reveal the crucial role of policy and regulations in mitigating the existing challenges. Insights from this research provide the construction industry managers with a comprehensive understanding of the major barriers to the adoption of passive cooling technologies and solutions to develop high-performance building envelopes.

Keywords Passive cooling · Green technology · High-performance envelope

1 Introduction

The world is warming faster than at any point in recorded history and the biggest challenge is keeping the indoor environments cool. According to the latest report of the United Nations Environment Program [51], the carbon emissions from the buildings' operation is in its all-time high of around 10GtCO2, which is 5% more than in

A. Fathalizadeh (✉) · E. Sharifi
School of Architecture and Civil Engineering, The University of Adelaide, Adelaide, SA, Australia
e-mail: ali.fathalizadeh@adelaide.edu.au

Z. Peng (ed.), *Technological Advancements and Future Directions in Green Energy*, Green Energy and Technology,
https://doi.org/10.1007/978-3-031-83203-1_9

2020, and more than 60% of this amount is demanded by the residential sector. Building envelopes are responsible for more than 50% of energy loss in the residential sector. Heating, ventilation, and air conditioning (HVAC) systems have been proved to account for 60% of the total energy consumed by buildings [3]. Various sustainable materials and cooling technologies have been introduced to reduce the force load on HVAC systems by enhancing the performance of the building envelopes.

Over the last few decades, there has been a surge in interest in what has become known as "high-performance building". The positive aspects of high-performance buildings include decreased environmental impact as well as economic and "value equation" gains from lower operating costs. Improved occupant comfort, security, and flexible space are some other advantages of high-performance passive envelopes [8, 20, 55]. Considering the emphasised impacts of global warming recently, the researchers' focus has mostly switched to the cooling aspect of the envelopes; defining a high-performance envelope as a building skin which keeps the interior of the building cool during the hot season and subsequently decreases the cooling load and energy demand by air conditioning systems. Maintaining a cool environment involves utilising a range of techniques and strategies, such as installing insulation systems to restrict heat transmission, installing shade and glazing systems to block out direct sunlight and regulate the inside temperature, and external surface coatings to keep the entire envelope cool. Despite the efforts taken by the researchers in making technological advancements and introducing innovative materials, their application in residential buildings is very limited. For instance, considering the case of Australia, it is proven that more than 80% of the existing residential buildings are not energy efficient. In the following, a review of the passive green technologies as well as the major challenges of their adoption is presented.

2 Background

2.1 High-Performance Passive Envelopes

Passive solutions are defined as any material, technology, or design feature adopted to provide indoor thermal comfort without energy consumption or power input. Focusing on the building envelopes, Sadineni et al. [43] review the envelope components where a passive technology can be applied. Different types of high-performance walls including Trombe walls, ventilated walls, and glazed walls as well as fenestration technologies like aerogel, vacuum glazing, and frames are discussed. High-performance roofs including contemporary green roofs, photovoltaic roofs, radiant-transmissive barrier, and evaporative roof cooling systems are also reviewed. Finally the thermal insulation systems, thermal mass, and phase change materials which can be used either in the walls or roof of the building envelope and airtightness and infiltration of the envelopes are investigated, providing a comprehensive view of the solutions available for a high-performance building envelope. Sadineni et al. [43] conclude that energy efficiency strategies may not necessarily require extra capital investment. An integrated energy-efficient building design can decrease the scale of mechanical systems, offsetting the added expense of energy-saving technologies.

Various studies have investigated the performance of building facades [8, 9, 37, 38, 47, 53] and presented diverse approaches to lower the energy consumption in buildings by improving the envelopes' performance. Different approaches to the design and construction of high-performance envelopes have been introduced and investigated through the literature. In the most conservative model introduced by Milardi [35], the window-to-wall ratio is in its lowest possible which prevents heat loss in winter and mitigates the heating effects due to direct solar radiation through the summer, sacrificing the occupant comfort with reducing the natural lighting. Another model, more suitable for hot and humid climates, utilises structural configurations to take advantage from the environmental conditions by allowing them inside the building. These include natural ventilation strategies, improved building layout, and building and windows orientation. The third model, as stated by Milardi [35], is named "regenerative installations" and relies on the systems to control the indoor environment and the role of the envelope is only as a barrier to limit the heat transfer between the interior and external. Studying the high-performance envelopes in cold climates of Canada, Aslani and Hachem-Vermette [8] prove that adopting high-performance walls in building envelopes can result in up to 80% reduction in energy consumption while having higher upfront environmental emissions compared with the conventional wooden stud walls. But considering the lifetime of the building (assumed as 50 years) CO_2 emission decreases equal to 231,189,938 kg. As presented in Table 1, passive technologies and solutions for the building envelopes are classified into four categories based on their application phase.

Table 1 Solutions for passive high-performance envelopes

Architectural design strategies	Building construction strategies	Materials		Add-on features and technologies
		Built-in materials	Retrofittable materials	
Window orientation	Insulation	Thermal mass	Phase change materials (PCM)	Shading
Top-window openings	Infiltration		Highly reflective materials	Vegetation
Façade cross ventilation	Air sealing		Glazing films	Radiative sky cooling (RSC)
Window-to-wall ratio	Double and triple pane glazing	Heat dissipation panel	External cladding systems	External blinds
Roof ponds	Curtain wall systems (CWS)		Microalgae	Thermal energy storage (TES)
Sunspace	Double skin façade (DSF)		Aerogel-Based Plasters	Timber shutters and lattices
References	[2, 5, 11, 14, 16, 17, 36, 39, 49, 52, 54]			

- The first category are the architectural and design strategies which should be considered in the early stages of building design. These include building orientation, window orientation, façade cross ventilation, and top-window openings. However, some of the design strategies can be incorporated into the building envelopes afterwards through a deep renovation process. For example, window-to-wall ratio, sunspaces, and roof ponds can be later added or modified in a building [12, 30].
- The second category refers to passive strategies which can be considered during the construction phase. Some examples include insulation, air sealing, double and triple pane glazing, and curtain wall systems.
- The next category refers to innovative green materials which can be chosen to replace the conventional ones in the building construction. The most popular ones are phase change materials (PCMs), highly reflective materials, thermal mass, glazing films, and aerogel-based plasters.
- The last category is the technologies and solutions which can be added to the buildings as an add-on feature every time. These are the most straightforward strategies for the building envelope retrofits. Shading solutions, vegetation, radiative sky cooling, and timber shutters and lattices are some examples in this category.

2.2 *Passive Technologies in Application Phase*

Application of various innovative solutions to building a high-performance envelope have been investigated by researchers in both academia and industry. In an industry driven research project, Lee et al. [32] focus on the role of Daylighting Technologies, Angular-Selective Shading Systems, and Dynamic Façades as three main pillars of a high-performance building envelope. Selkowitz [46] mentions the glazing and fenestration as the prominent elements of architectural form and emphasises the role of innovative glass technologies in moving the glass facades from energy losers to energy managers and ultimately energy suppliers. In another research work, Lee et al. [31] explore the growing interest in integrating advanced glazing components into commercial buildings' façade systems in a way that improves occupant comfort, productivity, and amenity, lowers operating costs for building owners, and contributes to environmental sustainability by using less energy and having fewer negative effects. Antoniadou et al. [6] compare conventional building facades which are developed over the time based on the geographical position and climate characteristics of each region with ventilated and double skin facades (as an example of high-performance envelopes) which are mostly introduced after 1990s and include more innovative passive and active technologies. While conventional facades are mostly solid walls built from layers of masonry materials, concrete, stone, or brick, the ventilated facades constitute an air space within the exterior and interior surface resulting in more efficient energy consumption. Notwithstanding the importance of ventilation and air flow in high-performance envelopes, the airtightness of the envelope is a major influence factor on the building energy performance [21]. Hachem-Vermette [28] concludes the criteria that a high-performance envelope should meet as preventing excess or undesirable heat gain while decreasing heat loss, air infiltration, and solar radiation control and management for use in passive heating and daylighting.

While most of the researchers studying high-performance envelopes are focused on optimising walls and windows, the roofs which play a significant role in the energy performance of a building remain neglected. The concept of "cool roofs" is the most popular research stream among the researchers. The cool roofs are well-known for their efficiency in reducing the building cooling demand through hot seasons by decreasing the roof surface temperatures [44, 50]. But their overall energy performance in cold climates is still debatable [50]. The finding of the study of (Tzempelikos and Lee) suggests that in warm climates the reflectivity of roofs has equal effect as the insulation of roofs, and for low-rise buildings reflective roofs offer higher savings than dark-coloured roofs. But in the case of cold climates dark-coloured roofs perform better than the reflective ones, especially for big-box-retail buildings.

2.3 Challenges Identified in the Literature

Several challenges and barriers to the passive cooling technologies and solutions have been identified by the researchers. These are either the barriers related to the technical nature of the solutions themselves, like the overheating in summer for Trombe walls and sunspaces [22], and high flammability for the organic PCMs [34, 40], or the external factors related to the policies and standards, public knowledge, and economic barriers, as illustrated in Fig. 1.

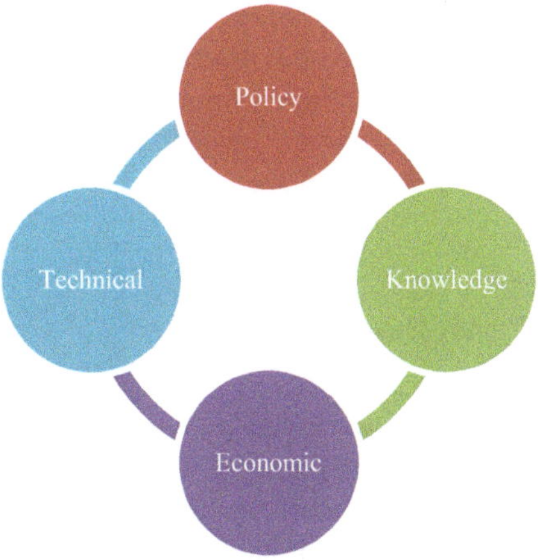

Fig. 1 Major barriers to the high-performance passive envelopes

Technical barriers are those related to the technological deficiencies. According to Sarihi et al. [45], bio-based materials, for example, have a number of drawbacks, such as inconsistent heat performance and condensation issues. Moisture can eventually lead to insulation deterioration, decay, mould growth, and loss of efficacy. Reflective coatings deteriorate more quickly than paints owing to soiling and weathering processes, which can lead to glare or excess incoming radiation, particularly in public areas. Additionally, while shading solutions are very effective in reducing indoor temperatures, a big limitation is that shading does not give occupants the ability to regulate glare. Static façade systems simply cannot always provide the optimal envelope response to the immediate environmental conditions [32], their cooling capacity is limited [54] and fail to fully meet energy demands.

Policy challenges and the barriers related to the standards and codes are the most controversial factors. While the researchers are able to conduct more research and develop innovative materials and technologies addressing the technical barriers, the policies cannot be modified easily considering the complicated factors involved. While Su et al. [48] and Hong et al. [30] find the obstacles is the lack of enough governmental incentives for passive cooling solutions. Friess and Rakhshan [26] believe in the need for more strict regulations and sanctions from the government. Lack of sufficient regional regulation and standardisation from the government, as well as the complexities in obtaining the renovation permits and the codes being ambiguous in this area have been identified as policy barriers [1, 7, 19].

Hong et al. [30] emphasise the lack of knowledge in various aspects in hindering the successful application of passive cooling solutions. The uncertainty in the cost benefits due to lack of enough research in different regions studying the cost-benefits and payback periods of high-performance passive envelopes as well as educating the public about the benefits is a big challenge with every innovative solution including high-performance envelopes. Enhancing the general knowledge of green façade technologies and providing realistic performance and cost data is what Hong et al. [30] suggest to facilitate the transition. Challenges with the lack of educated workforces and limited knowledge about the mode of execution are some more barriers related to the knowledge category [18, 33].

Adopting new technologies are always associated with economic and financial challenges. The high investments needed to build high-performance envelopes and additional costs for retrofitting envelopes with passive cooling technologies are considered as economic challenges hindering the people from moving towards high-performance envelopes [7, 13, 33, 48].

3 Methodology

In this study we aimed at first, understanding the passive technologies and solutions available in the literature for building high-performance envelopes. Subsequently, the challenges with adopting them have been identified through a comprehensive literature review. Finally, through interviews with the experts and professional in construction industry holding experience working with passive cooling solutions the significance of each major barrier has been revealed for the case of Australia. Through a purposive sampling method, experts who had more than 2 years of experience of working with passive cooling solutions in Australian residential sector were identified and contacted. Finally, 10 experts from different backgrounds, including the passive cooling materials supplier, building designers, strategic planners, and academic researchers, have been interviewed. The semi-structured interviews each lasting approximately 45 minutes, explored their experiences with passive cooling solutions and attitudes towards the challenges. Figure 2 shows the background distribution of the participants.

The methodology employed in this study involves the use of ordinal ranking and weighted scoring to assess the relative importance of major barriers to the adoption of passive envelope cooling solutions in the Australian residential sector. Ordinal ranking was chosen as the preferred method for several reasons. It offers a structured approach for capturing diverse participant perspectives and quantifying subjective opinions. Additionally, ordinal ranking is well-suited for survey-based research with ranked responses, such as our study on barrier categories. This method enables straightforward data analysis and comparison of relative importance among barriers, aiding in the identification of priority areas for intervention. Given the exploratory nature of our study, ordinal ranking provided an efficient means of prioritising barriers for further investigation and resource allocation [42].

In addition to the broader interview protocol, participants in this study were specifically queried on their perceptions of major barriers to the adoption of passive envelope cooling solutions within the Australian residential sector. This focused inquiry constituted a pivotal component of the comprehensive interview process and served as the basis for the analysis presented in this paper. Participants were tasked with ranking major barrier categories—policy, knowledge, technical, and economic—according to their perceived significance. Subsequently, each ranked factor was assigned a weight value, with the first-ranked factor assigned a value of 4, the second-ranked factor assigned 3, and so forth, down to the lowest-ranked factor, which was assigned a value of 1. This approach allowed for the quantification of perceived barriers and facilitated the identification of the most significant challenges in implementing passive cooling solutions.

Fig. 2 Interviewees' background distribution

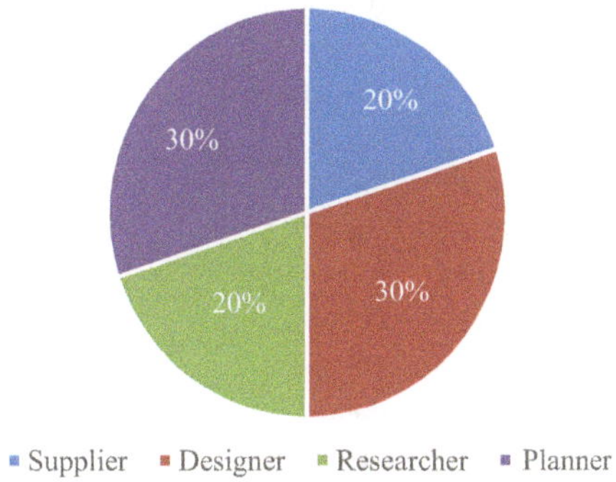

Relative Importance Index (RII), as shown in Eq. (1), is then used, adhering to the lessons of Holt [29], to rank the major barriers and reveal the most significant ones. This methodology have been applied by several researchers in similar field [23].

$$RII = 1n1 + 2n2 + \ldots + AnAAN \tag{1}$$

Where: $n1$, $n2$, … nA—number of interviewees scoring response stem integers 1 to 4; A—largest integer on the response scale; N—number of interviewees sample.
The A is equal to 4 in this research and $N = 10$.

4 Results and Discussion

Table 2 presents how each interviewee ranked the major barriers. Subsequently, each major barrier's RII was calculated, and the results are presented in Table 3. This shows that the interviewees ranked the "Policy" category as the most important one with an RII = 0.925. The second and third ranks go to the "Knowledge" with RII = 0.650, and "Economic" with RII = 0.625 respectively. The lowest rank belongs to the "Technical" barriers with RII = 0.3.

Analysis results reveal that the policy related challenges are by far the most significant barriers to adopting high-performance envelopes. This means that more support is needed from the government by providing initiatives, incentives, and subsidies facilitating the development of high-performance passive envelopes. This is also suggested by the interviewees that the force and sanction mechanisms in case of non-compliance with energy codes and standards would be effective to initiate the transition. The knowledge and economic barriers closely ranked second and third. The knowledge category includes both the public awareness and experienced workforces which are both needed for overcoming the barriers. The lowest ranked barriers were the technical challenges which most of the interviewees believed they

Table 2 Major barriers' ranking by interviewees

Interviewee ID	First rank (weight value = 4)	Second rank (weight value = 3)	Third rank (weight value = 2)	Forth rank (weight value = 1)
Interviewee 1	Policy	Knowledge	Technical	Economic
Interviewee 2	Knowledge	Policy	Economic	Technical
Interviewee 3	Policy	Knowledge	Technical	Economic
Interviewee 4	Policy	Economic	Knowledge	Technical
Interviewee 5	Policy	Economic	Knowledge	Technical
Interviewee 6	Knowledge	Policy	Economic	Technical
Interviewee 7	Policy	Economic	Knowledge	Technical
Interviewee 8	Policy	Economic	Knowledge	Technical
Interviewee 9	Policy	Economic	Knowledge	Technical
Interviewee 10	Economic	Policy	Knowledge	Technical

Table 3 RII values for each major barrier

Rank	Major barrier	RII
1	Policy	0.925
2	Knowledge	0.650
3	Economic	0.625
4	Technical	0.300

are not even a barrier, and a required level of technological advancement is already available.

Several researchers aimed at facilitating the transition towards sustainability in construction industry either by investigating the success factors [10] or failure factors [4, 24, 25, 41]. Although significant progress has been made in this area, the majority of earlier research has examined the bigger picture from a holistic perspective with limited practical solutions presented. More specific research studies on passive cooling solutions are also available [15, 27, 41], but they put a great emphasis on the technological aspects of the high-performance envelope development by introducing innovative materials and technologies and their performance assessment. Whereas the result of this study shows that supportive policies and stakeholders' knowledge are the most important barriers impeding the applications of high-performance building envelope.

5 Conclusion

This study aims at facilitating the transition towards green buildings through the adoption of passive cooling technologies. An overview of the high-performance passive envelopes and their challenges in the Australian residential sector have been presented in this study. The comprehensive literature review outlines the passive cooling solutions contributing to the construction of high-performance envelopes as well as the major barriers to their adoption. The challenges have been classified into major barriers of Technical, Policy, Knowledge, and Economic, which have been assessed in terms of relative importance through interviews with experts and professionals working with passive cooling technologies in the Australian construction industry. The participants of this study ranked the policy barriers as the most significant challenges hindering the adoption of high-performance envelopes. The knowledge and economic barriers closely ranked second and third and considering the technological advancements available, a majority of the interviewees did not consider technical issues as a barrier. Identifying the most influential barriers of passive cooling technologies offers invaluable insight for the managers in construction industry to better understand the obstacles and work through mitigating them. This also helps researchers, practitioners, and policy makers, in supporting the transition of residential buildings towards adopting green technologies and fulfilling sustainability goals.

While this research contributes valuable insights into the barriers to adopting passive envelope cooling solutions in the Australian context, several limitations should be acknowledged. Firstly, the geographical distribution of participants was confined to Australia, limiting the generalisability of the findings to other regions. Future studies should explore similar investigations in diverse geographical locations to compare results and offer more nuanced insights. Secondly, despite efforts to diversify participant backgrounds, the relatively small sample size of interviews conducted may introduce bias and limit the representativeness of perspectives. Additional interviews with a larger and more diverse sample could provide a broader understanding of the challenges faced. Lastly, the complex interrelation between identified barriers suggests that their influence may not follow a simple linear relationship. Further research utilising advanced analytical methods is warranted to explore the nuanced interactions between barriers and identify the most significant factors impacting the adoption of passive cooling solutions. By addressing these limitations, future research endeavours can enhance the robustness and applicability of findings in informing green building practices.

References

1. ABDUL MUJEEBU, M. & ALSHAMRANI, O. S. 2016. Prospects of energy conservation and management in buildings - The Saudi Arabian scenario versus global trends. *Renewable and Sustainable Energy Reviews,* 58**,** 1647–1663.
2. AFLAKI, A., MAHYUDDIN, N., AL-CHEIKH MAHMOUD, Z. & BAHARUM, M. R. 2015. A review on natural ventilation applications through building façade components and ventilation openings in tropical climates. *Energy Build.,* 101**,** 153–162.
3. AKEIBER, H., NEJAT, P., MAJID, M. Z. A., WAHID, M. A., JOMEHZADEH, F., ZEYNALI FAMILEH, I., CALAUTIT, J. K., HUGHES, B. R. & ZAKI, S. A. 2016. A review on phase change material (PCM) for sustainable passive cooling in building envelopes. *Renewable and Sustainable Energy Reviews,* 60**,** 1470–1497.
4. ALHAZMI, M., SAILOR, D. J. & LEVINSON, R. 2023. A review of challenges, barriers, and opportunities for large-scale deployment of cool surfaces. *Energy Policy,* 180**,** 113657.
5. AMIRIFARD, F., SHARIF, S. A. & NASIRI, F. 2019. Application of passive measures for energy conservation in buildings - a review. *Advances in Building Energy Research,* 13**,** 282–315.
6. ANTONIADOU, P., SYMEONIDOU, M., KYRIAKI, E., GIAMA, E., BOEMI, S. N., CHADIARAKOU, S. & PAPADOPOULOS, A. M. High performance building façades for Zero Energy Buildings in Greece: State of the art and perspectives. 2020. Institute of Physics Publishing.
7. ASCIONE, F., DE MASI, R. F., MASTELLONE, M., RUGGIERO, S. & VANOLI, G. P. 2020. Green Walls, a Critical Review: Knowledge Gaps, Design Parameters, Thermal Performances and Multi-Criteria Design Approaches. *ENERGIES,* 13.
8. ASLANI, A. & HACHEM-VERMETTE, C. 2022. Energy and environmental assessment of high-performance building envelope in cold climate. *Energy and Buildings,* 260.
9. BAEK, E., POHORYLES, D. A., KALLIORAS, S., BOURNAS, D. A., CHOI, H. & KIM, T. 2022. Innovative seismic and energy retrofitting of wall envelopes using prefabricated textile-reinforced concrete panels with an embedded capillary tube system. *Engineering Structures,* 265.

10. BANIHASHEMI, S., HOSSEINI, M. R., GOLIZADEH, H. & SANKARAN, S. 2017. Critical success factors (CSFs) for integration of sustainability into construction project management practices in developing countries. *International journal of project management,* 35, 1103–1119.
11. BOAFO, F. E., KIM, J. H., AHN, J. G., KIM, S. M. & KIM, J. T. 2021. Slim curtain wall spandrel integrated with vacuum insulation panel: A state-of-the-art review and future opportunities. *J. Build. Eng.,* 42.
12. BOUGIATIOTI, F., ALEXANDROU, E. & KATSAROS, M. 2023. Sustainable refurbishment of existing, typical single-family residential buildings in Greece. *International Journal of Building Pathology and Adaptation.*
13. CAO, X., DAI, X. & LIU, J. 2016. Building energy-consumption status worldwide and the state-of-the-art technologies for zero-energy buildings during the past decade. *Energy Build.,* 128, 198–213.
14. CHAI, J. & FAN, J. 2022. Advanced thermal regulating materials and systems for energy saving and thermal comfort in buildings. *Mater. Today Energy,* 24.
15. CHEN, X., YANG, H. & LU, L. 2015. A comprehensive review on passive design approaches in green building rating tools. *Renewable and Sustainable Energy Reviews,* 50, 1425–1436.
16. CHEN, X., YANG, H. & WANG, Y. 2017. Parametric study of passive design strategies for high-rise residential buildings in hot and humid climates: miscellaneous impact factors. *Renewable Sustainable Energy Rev,* 69, 442–460.
17. CRAWFORD, R. H., BARTAK, E. L., STEPHAN, A. & JENSEN, C. A. 2016. Evaluating the life cycle energy benefits of energy efficiency regulations for buildings. *Renewable and Sustainable Energy Reviews,* 63, 435–451.
18. CUCE, E. & RIFFAT, S. B. 2015. A state-of-the-art review on innovative glazing technologies. *Renewable and Sustainable Energy Reviews,* 41, 695–714.
19. DADZIE, J., RUNESON, G. & DING, G. 2020. Assessing determinants of sustainable upgrade of existing buildings: The case of sustainable technologies for energy efficiency. *J. Eng. Des. Technol.,* 18, 270–292.
20. ENGEBØ, A., KLAKEGG, O. J., LOHNE, J. & LÆDRE, O. 2020. A collaborative project delivery method for design of a high-performance building. *International Journal of Managing Projects in Business,* 13, 1141–1165.
21. ERHORN-KLUTTIG, H., ERHORN, H., LAHMIDI, H. & ANDERSON, R. 2009. Airtightness requirements for high performance building envelopes.
22. FARAJ, K., KHALED, M., FARAJ, J., HACHEM, F. & CASTELAIN, C. 2020. Phase change material thermal energy storage systems for cooling applications in buildings: A review. *RENEWABLE & SUSTAINABLE ENERGY REVIEWS,* 119.
23. FATHALIZADEH, A., GHODDOUSI, P., SHIRZADI JAVID, A. A., HOSSEINI, M. R. & GHOSH, A. 2019. Integrating sustainability into construction project management: Barriers in developing countries. *13th International Conference; MODERN BUILDING MATERIALS, STRUCTURES AND TECHNIQUES.* Vilnius, Lithuania.
24. FATHALIZADEH, A., HOSSEINI, M. R., SILVIUS, A. J. G., RAHIMIAN, A., MARTEK, I. & EDWARDS, D. J. 2021a. Barriers impeding sustainable project management: A Social Network Analysis of the Iranian construction sector. *Journal of Cleaner Production,* 318, 128405.
25. FATHALIZADEH, A., HOSSEINI, M. R., VAEZZADEH, S. S., EDWARDS, D. J., MARTEK, I. & SHOOSHTARIAN, S. 2021b. Barriers to sustainable construction project management: the case of Iran. *Smart and Sustainable Built Environment,* ahead-of-print.
26. FRIESS, W. A. & RAKHSHAN, K. 2017. A review of passive envelope measures for improved building energy efficiency in the UAE. *Renewable Sustainable Energy Rev,* 72, 485–496.
27. GASSAR, A. A. A., KOO, C., KIM, T. W. & CHA, S. H. 2021. Performance optimization studies on heating, cooling and lighting energy systems of buildings during the design stage: A review. *Sustainability,* 13.
28. HACHEM-VERMETTE, C. 2020. *Solar Buildings and Neighborhoods; Design Considerations for High Energy Performance*, Springer Cham.

29. HOLT, G. D. 2014. Asking questions, analysing answers: relative importance revisited. *Construction Innovation,* 14, 2–16.
30. HONG, W. T., IBRAHIM, K. & LOO, S. C. 2019. Urging green retrofits of building facades in the tropics: A review and research agenda. *Int. J. Technol.,* 10, 1140–1149.
31. LEE, E., SELKOWITZ, S., BAZJANAC, V., INKAROJRIT, V. & KOHLER, C. 2002. High-performance commercial building facade.
32. LEE, E. S., COFFEY, B., FERNANDES, L., HOFFMAN, S., MCNEIL, A., THANACHAREONKIT, A. & WARD, G. 2014. *High Performance Building Façade Solutions-Phase II,* Lawrence Berkeley National Laboratory. LBNL Report #: LBNL-1004337.
33. M'LAHFI, B., AMEGOUZ, D. & EL QANDIL, M. 2020. A new approach for the mandatory application of the thermal regulation of construction (RTCM) in the future moroccan buildings. *SN Appl. Sci.,* 2.
34. MA, Q., CHEN, X., WANG, X., GAO, W., WEI, X. & FUKUDA, H. 2021. A review of the application of sunspace in buildings. *Energy Sources Recovery Util. Environ. Eff.*
35. MILARDI, M. 2021. Integrated, Adaptive and Smart Envelope for Near Zero Energy Buildings. *New Metropolitan Perspectives.*
36. MIRRAHIMI, S., MOHAMED, M. F., HAW, L. C., IBRAHIM, N. L. N., YUSOFF, W. F. M. & AFLAKI, A. 2016. The effect of building envelope on the thermal comfort and energy saving for high-rise buildings in hot-humid climate. *Renewable and Sustainable Energy Reviews,* 53, 1508–1519.
37. MOGHTADERNEJAD, S., CHOUINARD, L. E. & MIRZA, M. S. 2018. Multi-criteria decision-making methods for preliminary design of sustainable facades. *Journal of Building Engineering,* 19, 181–190.
38. MOGHTADERNEJAD, S., CHOUINARD, L. E. & MIRZA, M. S. 2020. Design strategies using multi-criteria decision-making tools to enhance the performance of building façades. *Journal of Building Engineering,* 30.
39. OZARISOY, B. & ALTAN, H. 2021. Systematic literature review of bioclimatic design elements: Theories, methodologies and cases in the South-eastern Mediterranean climate. *Energy Build.,* 250.
40. PODARA, C. V., KARTSONAKIS, I. A. & CHARITIDIS, C. A. 2021. Towards phase change materials for thermal energy storage: classification, improvements and applications in the building sector. *Appl. Sci.,* 11, 1–26.
41. POMBO, O., RIVELA, B. & NEILA, J. 2016. The challenge of sustainable building renovation: Assessment of current criteria and future outlook. *Journal of Cleaner Production,* 123, 88–100.
42. ROSZKOWSKA, E. 2013. Rank ordering criteria weighting methods–a comparative overview. *Optimum. Studia Ekonomiczne*, 14–33.
43. SADINENI, S. B., MADALA, S. & BOEHM, R. F. 2011. Passive building energy savings: A review of building envelope components. *Renewable and Sustainable Energy Reviews,* 15, 3617–3631.
44. SANTAMOURIS, M. & ASIMAKOPOULOS, D. 1996. *Passive Cooling of Buildings*, James & James.
45. SARIHI, S., MEHDIZADEH SARADJ, F. & FAIZI, M. 2021. A Critical Review of Façade Retrofit Measures for Minimizing Heating and Cooling Demand in Existing Buildings. *Sustainable Cities Soc.,* 64.
46. SELKOWITZ, S. E. 2001. *Integrating advanced facades into high performance buildings,* Lawrence Berkeley National Laboratory.
47. SIMOES, N., GONCALVES, M., SERRA, C. & RESALATI, S. 2021. Can vacuum insulation panels be cost-effective when applied in building facades? *Building and Environment,* 191, 15.
48. SU, Y., WANG, L., FENG, W. & ZHOU, N. 2021. Analysis of green building performance in cold coastal climates: An in-depth evaluation of green buildings in Dalian, China. *Renewable Sustainable Energy Rev,* 146.

49. TALAEI, M., MAHDAVINEJAD, M. & AZARI, R. 2020. Thermal and energy performance of algae bioreactive façades: A review. *J. Build. Eng.*, 28.
50. TZEMPELIKOS, A. & LEE, S. 2021. Cool Roofs in the US: The Impact of Roof Reflectivity, Insulation and Attachment Method on Annual Energy Cost. *Energies*, 14.
51. UNEP 2022. 2022 Global Status Report for Buildings and Construction: Towards a Zero-emission, Efficient and Resilient Buildings and Construction Sector. Nairobi: United Nations Environment Programme.
52. VALLADARES-RENDÓN, L. G., SCHMID, G. & LO, S. L. 2017. Review on energy savings by solar control techniques and optimal building orientation for the strategic placement of façade shading systems. *Energy Build.*, 140, 458–479.
53. WILLIAM, M. A., SUÁREZ-LÓPEZ, M. J., SOUTULLO, S. & HANAFY, A. A. 2021. Building envelopes toward energy-efficient buildings: A balanced multi-approach decision making. *International Journal of Energy Research*, 45, 21096–21113.
54. WU, Y., ZHAO, H., SUN, H., DUAN, M., LIN, B. & WU, S. 2022. A review of the application of radiative sky cooling in buildings: Challenges and optimization. *Energy Convers. Manage.*, 265.
55. ZERKIN, A. J. 2006. Mainstreaming high performance building in New York City: A comprehensive roadmap for removing the barriers. *Technology in Society*, 28, 137–155.

Index

Z. Peng (ed.), *Technological Advancements and Future Directions in Green Energy*, Green Energy and Technology,
https://doi.org/10.1007/978-3-031-83203-1

GPSR Compliance
The European Union's (EU) General Product Safety Regulation (GPSR) is a set of rules that requires consumer products to be safe and our obligations to ensure this.

If you have any concerns about our products, you can contact us on

ProductSafety@springernature.com

In case Publisher is established outside the EU, the EU authorized representative is:

Springer Nature Customer Service Center GmbH
Europaplatz 3
69115 Heidelberg, Germany

www.ingramcontent.com/pod-product-compliance
Ingram Content Group UK Ltd.
Pitfield, Milton Keynes, MK11 3LW, UK
UKHW021409091026
15624UKWH00009B/85
9783031832024